MEDIA MANIA

Hugh Mackay is a psychologist, social researcher, columnist and novelist.

He was educated at Sydney Grammar School, the University of Sydney and Macquarie University, and established Mackay Research in 1971. Since 1979, he has been publishing regular reports based on a continuous program of qualitative research into the changing attitudes and behaviour of the Australian community. In recognition of his pioneering work in social research, he has been awarded honorary doctorates by the University of New South Wales, Macquarie University and Charles Sturt University.

He is a Fellow of the Australian Psychological Society, an honorary professor in the Macquarie Graduate School of Management, Chairman of Trustees of Sydney Grammar School and a former Deputy Chairman of the Australia Council.

He writes a weekly newspaper column on social issues for the *Sydney Morning Herald*, the *Age* and the *West Australian* and is the author of four bestsellers in the field of social psychology. His fourth novel, *Winter Close*, was published in 2002.

Other books by Hugh Mackay

Non-fiction
Reinventing Australia
The Good Listener
Generations
Turning Point

Fiction
Little Lies
House Guest
The Spin
Winter Close

MEDIA MANIA

WHY OUR FEAR OF MODERN MEDIA IS MISPLACED

Hugh Mackay

A UNSW Press book

Published by
University of New South Wales Press Ltd
University of New South Wales
UNSW Sydney NSW 2052
AUSTRALIA
www.unswpress.com.au

© The Trustees of the New College Lectures 2002
First published 2002

This book is copyright. Apart from any fair dealing for the purpose of private study, research, criticism or review, as permitted under the Copyright Act, no part may be reproduced by any process without written permission. Inquiries should be addressed to the publisher.

National Library of Australia
Cataloguing-in-Publication entry:

   Mackay, Hugh, 1938– .
   Media mania: why our fear of modern media is misplaced.

   Bibliography.
   ISBN 0 86840 709 7.

   1. Mass media — Influence. 2. Mass media — Social aspects.
   I. Title. (Series: New College lectures).

   302.23

Printer Southwood Press

NEW COLLEGE LECTURES AND PUBLICATIONS

New College is an Anglican college affiliated with the University of New South Wales, Sydney. In 1986 the college set up a trust to conduct an annual series of public lectures. The lecturer is asked to take up some aspect of contemporary society and to comment on it from the standpoint of their Christian faith and professional experience. The inaugural lectures were given in 1987 by Professor Malcolm Jeeves of the University of St Andrews, Scotland, and in subsequent years lecturers have come from Australian and overseas universities, as well as the wider community.

1987 Prof Malcolm Jeeves (University of St Andrews, Scotland)
 Minefields, Lancer Books (ANZEA), 1994.
1988 Veronica Brady (University of Western Australia)
 Can These Bones Live? Federation Press, 1997.
1989 The Hon Justice Keith Mason (NSW Supreme Court)
 Constancy and Change, Federation Press, 1990.
1990 Prof Stanley Hauerwas (Duke University, USA)
 After Christendom? ANZEA, 1991.
1991 Prof Geoffrey Bolton (University of Queensland).
1992 Prof Peter Newman (Murdoch University, Western Australia).

1993 Prof Robin Gill (University of Kent, England)
Beyond Self Interest, New College, 1993.
Rev Dr John Polkinghorne, KBE, FRS (Queens' College, Cambridge, England)
Religion and Current Science, New College, 1993.
1994 Prof Geoffrey Brennan (Australian National University, Canberra).
1995 Rev Dr John Polkinghorne, KBE, FRS (Queens' College, Cambridge, England)
Beyond Science, Cambridge University Press, 1996; Polish ed, 1998; Greek ed, 1999.
1996 Les Murray (Australian poet)
Killing the Black Dog, Federation Press, 1997.
1997 Dr Elaine Storkey (London Institute for Contemporary Christianity, England)
Created or Constructed? The Great Gender Debate, Paternoster Press, 2000; UNSW Press, 2001.
1998 Dr Peter Vardy (Heythrop College, University of London, England)
What is Truth? UNSW Press, 1999.
1999 Prof William J Baker (University of Maine, USA)
If Christ Came to the Olympics, UNSW Press, 2000.
2000 Prof Hilary Charlesworth (Australian National University, Canberra)
Writing in Rights: Australia and the Protection of Human Rights, UNSW Press, 2001.
2001 Hugh Mackay (Mackay Research, Sydney)
Media Mania: Why our Fear of Modern Media is Misplaced, UNSW Press, 2002.

CONTENTS

	Foreword by Allan Beavis	9
	Introduction	13
1	The television bug	17
2	What kind of violence?	39
3	'I'll just check my email … again'	59
	Postscript: The 'Big Brother' phenomenon	79
	Bibliography	83

FOREWORD

by Allan Beavis
Master, New College UNSW

Hugh Mackay is a gifted communicator. His 2001 New College Lectures were a treat in terms of their eloquence and elegance. Hugh's engaging style made the contents of his lectures accessible and even entertaining. It has been a great pleasure to have the opportunity to revisit the lectures in reading the manuscript of this book. Here too, Hugh's communications skills are amply demonstrated. The nuances of his personal delivery may be lost in the written text, but they are adequately compensated for in the charm of his highly appealing writing style. This is an engaging book with much to offer.

The New College Lectures are an annual series of public lectures delivered in New College on the campus of UNSW. They provide an opportunity for an eminent scholar or practitioner to take up some aspect of contemporary society and comment upon it from the standpoint of their own Christian faith and professional experience. Some time ago, the Trustees of the Lectures concluded that the place of the rapidly changing and highly technologised media in contemporary society was a topic in need of closer scrutiny. After careful consideration, Hugh Mackay, social researcher and commentator, was the person identified as being best able to address this topic in an Australian context.

In his three lectures (which form the bases for the three chapters in this book), Hugh provided a penetrating analysis of contemporary and evolving forms of mass media, derived from his own research. He convincingly countered the prevailing wisdom of the media's inflated power to influence attitudes and behaviours, pointing rather to its ability to reinforce attitudes and behaviours that are already extant and influenced by other more powerful social institutions. He put into perspective assertions about the causal links between media violence and violent behaviour and suggested that the real violence perpetrated by the media is cultural and located in the prevalent sub-text that 'everything should be entertaining'. Finally, he analysed the nature of communication itself, drawing attention to its highly personal character, requiring as it does the presence of two communicating human beings. He discussed the extent to which contemporary media can facilitate genuine communication or whether, more commonly, they merely mediate the transference of information and data.

The notion of communication and how it is mediated are concepts neither clearly understood nor agreed upon in academia or society at large. The most common understanding of communication would be of sending and receiving messages. In this paradigm a communicator sends a message to an addressee who (ideally) receives the message as sent almost as if something tangible has been transmitted from sender to receiver. An alternative, but perhaps less accepted, understanding of communication is of two information processing units disturbing one another's internal states and in the process effecting communication. However communication may be viewed, it is nevertheless an essential social process. Indeed, it is arguably *the necessity* of social life. Without communication there can be no such thing as society. How communication is mediated is therefore a matter of singular social importance.

Essentially this is an optimistic book. It demythologises many prevailing claims concerning communications media that have generated fear and anxiety — especially among the generation that grew up in the pre-TV, pre-Internet and pre-mobile-phone era.

While it is realistic in drawing attention to problems that can justly be attributed to the media (such as the shortening of our attention span) it also points to how contemporary media can be used to serve what is so essential to our humanity — our need to maintain interpersonal relationships within community.

The New College Lectures have a proud history and the list of previous lecturers and publications is impressive. The lectures are delivered within the academic context of the University of New South Wales, and the lecturers are given freedom to handle the agreed topic as they choose. The views they express are not necessarily those of the College.

The Trustees thank Hugh Mackay for a splendid series of lectures and a publication that enhances the contribution of this series to Australian society. They are also grateful to Robin Derricourt, Nicola Young and Selina Altomonte from UNSW Press for their assistance in publication.

INTRODUCTION

People who feel disgruntled about the state of contemporary society are always on the lookout for a scapegoat: it might be the rate of family breakdown, the lack of visionary leadership, the decline of religion, the introduction of the contraceptive pill, or the impact of economic rationalism. But whenever I speak to audiences about the phenomenon of social change, someone will usually stand up during question-time and declare that I've overlooked the most obvious cause of all our troubles: the mass media, especially television.

'Wouldn't you agree', they often begin, 'that television has made us a more violent society?'. Sometimes the charge is more general: 'Wouldn't you say that television is responsible for our moral decline — that the media are teaching our children an undesirable set of values?'.

Such questions sadden me, for two reasons. First, they mistakenly assume that the mass media are a more powerful influence on us than we are on each other; that watching television is a more formative experience than the experience of living in a particular kind of family. Second, they betray a willingness — even an eagerness — to blame anyone or anything other than ourselves for the state of our society.

These two themes — how the media affect us, and how the media might contribute to the ills of society — permeate the chapters of this book. *Media Mania* began life as a series of three lectures — The New College Lectures — delivered at the University of New South Wales during October 2001. The brief from the Master of New College, Dr Allan Beavis, was to examine the role of the mass media in society, drawing on my own experience as a social researcher.

This is a subject close to my heart. My entire working life has been spent studying the attitudes and behaviour of the Australian community, and much of that study has focused on the role and function of the media. When I joined the audience research department of the Australian Broadcasting Commission, television had only been operating in Australia for four years. (Earlier, at The McNair Survey, I had worked on Australia's first TV ratings survey.) So we were fascinated by the potential of this amazing new medium and, like so many other researchers at the time, we were trying to fathom its likely impact on people who were devoting vast chunks of their lives to staring at the TV screen — even making appointments for their TV-deprived neighbours to come over and watch particular programs.

At the ABC, I worked for Dr Peter Kenny, a wildly eccentric but brilliant researcher who, in turn, introduced me to the pioneering work of Dr Joseph Klapper, head of the office of social research at the American CBS network. Klapper subsequently became a friend and mentor, and his work was a revelation to me: it was the first serious challenge to my naïve acceptance of the idea that television — or any medium — exerted a direct influence on us *via its program content*. How easy it was to believe (yet how wrong!) that if we watched a particular kind of program — crime drama, for instance — this would shape our attitudes and behaviour in ways that mimicked the program.

Marshall McLuhan also burst onto the scene at that time, declaring that *the medium itself* was the message. McLuhan argued that we had been looking in the wrong place for evidence of media effects: we should have been looking at the effect of reading the printed word, for instance, rather than the effect of reading particular books; we should have been looking at the impact of the TV viewing experience itself, rather than individual programs, if we wanted to understand what television was doing to us.

McLuhan has his critics, but his ideas re-shaped the approach of many researchers, causing us to look at the question of media effects in a broader and more fundamental way. Once we turned our backs on the 'injectionist' model of mass communication, we found the field to be more rewarding and far more challenging. Recently, for instance, we've had to explain the fact that violence in the United States has been steadily declining for 10 years, with no apparent diminution in the amount of violent material on television, at the cinema, or in video games. And we've had to examine the possibility that television works like a self-administered tranquilliser for millions of Australians who only ever dip into its offerings, rarely staying to watch an entire program.

This reworking of my New College Lectures is offered in the hope that it will encourage readers — especially parents — to adopt a more realistic view of the role and function of the media in our lives. We must fight the victim mentality here, as in so many departments of our lives, and acknowledge that the effects of television are far more dependent on what we do with the medium than on what the medium does with us.

There is a kind of 'media mania' that causes us to feel threatened by the media, as though they have some inherent power over us. The same mania leads us to the dangerous assumption that the media are shaping our society in ways we can neither understand nor control. My message to parents is simple: you have far more influence over your children's attitudes and values than television will ever have. To all of us, the central message is that while the media may change the *way* we think and communicate, we are still responsible for *what* we think and *what* we say. Neither the media, nor any other scapegoat, can absolve us of responsibility for the quality and structure of our own lives.

In presenting these three lectures (and, as a light-hearted postscript, the text of an informal talk I gave to the students of New College over dinner), I received encouragement, support and warm hospitality from Dr Beavis and his colleagues. I particularly enjoyed my contact with the students of New College, and it is to them that I dedicate this book.

1

THE TELEVISION BUG

I once toyed with an idea for a short story called *The Television Bug*, about a group of evil geniuses who were developing a weird kind of electronic virus that could be transmitted via TV broadcasts, thus infecting entire viewing audiences.

The idea was that these viruses would somehow carry with them the essence of messages embedded in the program content: when they were transmitted via commercials, people would fall in helplessly with the advertisers' recommended courses of action. Thanks to these bugs, political broadcasts would have a previously undreamed-of power to persuade. When the bugs were released during programs showing violence and mayhem, the population would be roused to new heights of aggression. If they were transmitted via soft-porn movies ... well, you get the idea.

(If I'd been as smart as Richard Dawkins (1989), I might have dubbed these viral television bugs 'tele-memes' ... but that's another story.)

In fact, it wasn't such a radical thought. It was simply an extension of a popular idea that runs deep in our culture: the idea that TV program content has the power to change the way we are — to shape our tastes, our preferences, our beliefs and our values and, even more disturbingly, to modify or even control our behaviour.

Though I imagine no one seriously believes that there is some *direct* channel of influence running from the TV screen to the mind of the viewer, both the harshest critics and the staunchest advocates of television seem to regard the medium almost as if it has that kind of inherent power ... or even as if they *wish* it had that kind of power, so the critics would have something substantial to attack and the defenders would have something worth defending.

Some zealous protectors of children's rights, for instance, talk about the effects of exposure to TV programs — especially violent programs — almost as though it will turn our children into a more aggressive generation of humans than we've ever seen before — Genghis Khan (a non-TV-viewer, as it happens) notwithstanding.

Many advertisers approach television as if it is a medium that fires magic bullets into the minds of its viewers, thus repaying their vast expenditure on TV commercials with spectacular changes in the behaviour of their unwitting consumers. ('The power of advertising' is an oft-heard phrase, but it should be used to refer to advertising's proven power to reinforce existing attitudes and preferences, rather than its power to convert.)

In fact, most of us seem to almost intuitively assume that television has the power to influence people like no other medium — personal or impersonal — can. The church is no exception: since the advent of television, church leaders have been flirting with the idea that, if only the medium could be properly harnessed, television could offer a dramatic new way to preach the gospel to millions of viewers who would never otherwise encounter it and who certainly have no intention of going to church.

When the Australian Government started to believe that the illicit drug trade was slipping too far beyond its control and that the so-called 'war on drugs' was not working, what did it do? It turned to the mass media for help: a printed brochure was distributed to every household in the nation, and Australia's TV screens were saturated with 'Tough On Drugs' messages. Someone, presumably, believed that the expenditure of millions of media dollars would produce a dramatic fall in Australians' — especially young Australians' — proclivity to use illicit drugs, though no such outcome has yet been announced.

It is an extraordinary feature of our society that we maintain such strong faith in the idea that television is peculiarly equipped with the power to transform us — for good or ill, but mostly ill — yet we willingly turn it on every night of the week and expose ourselves, hour after hour, to its seductions, blandishments and — for all I know — its electronic viruses.

Perhaps each of us believes that we are uniquely protected from the dark forces of the media. There does seem to be a widespread belief that, by some quirk of fate, we ourselves, as individuals, are somewhat immune to the impact of the television bug. One of the most common paradoxes in popular discussion of media effects is revealed when people complain bitterly about a media story — a TV program, say, or even a newspaper column — on the grounds that it represents 'brainwashing', yet simultaneously assert that they themselves are not merely unmoved by the assault but, indeed, are both wise to it and utterly resistant to its presumed effects on everyone else. (I couldn't count the number of letters I have received from angry readers in response to my own weekly newspaper column, saying things like: 'You're brainwashing the public with your bleeding-heart, soft-left politics — but you don't fool *me* for a moment. We can make up our *own* minds about who to vote for, thank you'. My point, precisely.)

Isn't it odd? We somehow manage to believe in the wondrous power of the media — especially television — to influence everyone but us. Somewhere out there in the community, there are apparently thousands upon thousands of vulnerable, innocent souls who are swept ceaselessly — this way and that — by every bit of media propaganda. But not me.

It's a view with a long and glorious tradition. There's a popular misconception about the way the mass media operate in our society that can best be described as 'the hypodermic effect'. The hypodermic view of mass communication treats media content rather like a drug and the media themselves as giant hypodermic syringes, capable of injecting a mass audience at a single stroke. In goes the drug — via the eye and ear — and the brain is irrevocably changed by the experience. Now we've been exposed to the latest episode of 'The Bill' we'll be looking for new and smarter ways to evade the long arm of the law. Now we've seen the latest commercial for Westpac we'll glide, as if sedated, into our local branch of the Commonwealth Bank, close all our accounts and stagger across to the road to Westpac carrying a sack of money — or, more likely, a sack of debt — until we see an equally persuasive commercial for the Commonwealth Bank and then, presumably, we'll stagger back across the street, repeating the whole process in reverse.

It's a long-standing view, popularised by such writers as Vance Packard (1957) and Marie Winn (1978). By now, though, we should be starting to become wary of the idea that televised messages are all-powerful: where are those magic advertising campaigns that were going to convert us all into brand-obedient automatons? Where are these religious broadcasts that were going to stimulate a new religious revival, or those election campaign commercials with the power to sway the electorate one way or the other?

Even if you don't have such a simplistic view of the hypodermic effect, it may still persist in a more refined form: you might feel that *it goes without saying* that children who are exposed to high doses of media violence will be stimulated to perform a correspondingly high number of aggressive acts. Surely, you might find yourself asking, when children have watched tens of thousands of violent acts on television before the age of five, won't this make them more aggressive than previous generations who were not exposed to such violent imagery?

Such intuitions are easy to understand, and there is plenty of research that appears to support them — though much of it is poor research, and there is as much or more research to challenge or even reject that hypothesis.

But let's try to be as rigorous as we can be about this process — given that perfect scientific rigour is never possible when we're dealing with something as complex, subtle and non-rational as human responses to media messages. Still, let's try: let's at least look at how the mass-communication process actually works, and the implications for the belief that there's a 'television bug' loose in the community, eating away at our brains ... and, indeed, our souls.

HOW DOES MASS COMMUNICATION WORK?

Forty years ago, research into the process and effects of mass communication took a significant turn. Traditionally, studies of mass-media effects (including the effects of advertising) were based on an implicit assumption that the media effect *was* 'hypodermic': that there *was* a kind of television bug that invaded people's minds and lodged there, doing its mischievous, furtive work. Traditionally, therefore, research into the effects of mass communication tended to concentrate on the recall of program content, searching for the 'residue' of the program material.

But 1960 was a landmark year in our approach to mass-communication research and, although that seems a long time ago now, one particular book contained one particular set of generalisations about the mass-communication process that have not only stood the test of time but have become the most cited work in the literature of mass communication. That book was *The Effects of Mass Communication* and its author was Joseph Klapper (1960). Klapper, daringly and provocatively, described five 'emerging generalisations' based on a new wave of research into the effects of mass communication. All these years later, and in spite of many theoretical and philosophical cross-currents, undertows and rips, the generalisations are still bravely standing. Here they are, exactly as Klapper proposed them:

1 Mass communication ordinarily does not serve as a necessary and sufficient cause of audience effects, but rather functions among and through a nexus of mediating factors and influences.

2 These mediating factors are such that they typically render mass communication a contributory agent, but not the sole cause, in a process of reinforcing the existing conditions.
3 On such occasions as mass communication does function in the service of change, one of two conditions is likely to exist. *Either:*
   (a) The mediating factors will be found to be inoperative and the effect of the media will be found to be direct; *or*
   (b) The mediating factors, which normally favour reinforcement, will be found to be themselves impelling toward change.
4 There are certain residual situations in which mass communication seems to produce direct effects, or directly and of itself to serve certain psycho-physical functions.
5 The efficacy of mass communication, either as a contributory agent or as an agent of direct effect, is affected by various aspects of the media and communications themselves or of the communication situation (including, for example, aspects of textual organisation, the nature of the source and medium, the existing climate of public opinion, and the like).

In case you think I am exaggerating when I say that Klapper's generalisations are still valid forty years on, a recent review by Richard Felson (1996) of the literature on the effects of exposure to media violence had this to say:

> The message that is learned from the media about when it is legitimate to use violence is not much different from the message learned from other sources ...

and this:

> It is not clear what lesson the media teaches about the legitimacy of violence, or the likelihood of punishment. To some extent that message is redundant with lessons learned from other sources of influence. The message is probably ambiguous and is likely to have different effects on different viewers. Young children may imitate illegitimate violence, if they do not understand the message, but their imitative behaviour may have trivial

consequences. Out of millions of viewers, there must be some with highly idiosyncratic interpretations of television content who intertwine the fantasy with their own lives ...

Felson's 1996 conclusions, examined more fully in the following chapter, clearly resonate with Klapper's original generalisations. The emphasis is on media effects as *contributory*, on the nexus of mediating factors through which media effects must pass, and on the possibility of direct effects only when there are no mediating factors (as, for example, in the case of young, inexperienced viewers of TV programs who have no relevant life experience and no countervailing messages coming from the real-world context of family or friends, or in the case of messages about a new subject about which the audience is ignorant and therefore something of a 'blank slate').

The essential truth about the process of mass communication is that its main effect is to reinforce the status quo. It tends to give the audience what they want. It tends to satisfy pre-existing needs in the audience for stimulation, for tranquillisation or for more specific reflection of existing attitudes, values, aspirations or beliefs.

How could it be otherwise? As viewers, we don't come unencumbered to the TV viewing experience: we bring our own expectations and prejudices with us and they have a profound effect on the quality and character of our encounter with the medium. They also make it easy to reject, ignore or deflect messages we don't like, and to perceive and interpret messages in a highly selective way.

Imagine every member of the media audience watching and listening from inside the protective shield of a personal, psychological shell, constructed out of all their accumulated learning. At an early age, each of us begins the lifelong work of building that protective shell, or cocoon: we strive to make sense of the world by weaving together all the threads of our learning — the fruits of our experience — into a coherent pattern. As we create these patterns to help us make sense of what's happening to us, we gradually develop a world view (which also acts, paradoxically, as a kind of filter between us and the world).

We need a comfortable and secure cocoon, or shell, for our sense of personal identity and for our emotional security. Stable attitudes and beliefs are part of what it means to be sane: they help us withstand attacks from those who disagree with us, or want to persuade us to adopt a different view of the world (based on *their* experience rather than our own).

But, from inside the cocoon of our beliefs and prejudices, we don't see the world in a direct, unfiltered way: we see it through the pattern created by those beliefs and prejudices. Our attitudes are integral with our perceptions, and we lay them — rather like a template — over our view of the world. This is particularly easy to do in the case of TV program content, because that is less 'real' to us than most of the experiences that have given rise to our attitudes and beliefs. (If there is a television bug, by the way, it has to sneak in through the tightly-wound fibres of our protective cocoons.)

This is why two viewers can view the same TV program and come away from the experience with utterly different impressions of what they have seen. It's why two people with diametrically opposed political views can watch a political campaign commercial, or hear the broadcast of a speech, and be roused to enthusiasm or fury, respectively.

The metaphor of the cocoon helps to explain why, when our point of view is attacked, we automatically defend it and why, in the process, our convictions are strengthened, not weakened. When religious minorities are persecuted, this generally has the effect of *reinforcing* their beliefs: you don't hear stories of persecuted minorities saying, in the face of hostility, 'Oh, we must have been wrong: let's give this up and believe something else'. On the contrary: nothing nurtures and strengthens faith like the need to defend it in the face of persecution.

Because they are in early stages of constructing their protective cocoons, children need special protection and guidance in their use of mass media. Parents, teachers and friends need to educate children about how to approach and interpret the mass media — just as we need to educate and guide them in their approach to every other kind of influence in the social, cultural and physical environment. Nevertheless, children weave their cocoons at a furious pace and, from an early age, seem well able to distinguish between programs and 'reality'. Our 1983 report on *Children and Television* came to this conclusion about 8–10-year-olds' attitudes to television:

> TV is like a game. An important distinction between 'real life' and games is that games produce a quick result, whereas life just goes on. In games, there are clearly-defined winners and losers, whereas in real life, there is good and bad in everyone and no automatic sense of justice. In games, more extreme forms of behaviour (cheekiness, violence, antagonism, shrieking and yelling, etc) are tolerated than would be tolerated in real life. In games, roles are adopted and played out which are often quite different from those adopted in real life.
>
> In such ways, television seems [to children] to be more of a game than a department of real life. The way children talk about their use of television and the gratifications which they obtain from television suggests that they do not often expect or want the medium to be about reality. TV is not 'real' in the way that home, family, friends, teachers and schools are real. TV is much more in the category of games.
>
> Indeed, when children talk about what they would do if they were not viewing TV, they invariably turn to activities like playing cards, having bike races, playing football, or 'just playing'.

Since then, video games have demonstrated the point rather literally. Nevertheless, Klapper's third generalisation is salutary: the mass media can have a direct impact on attitude formation where the normal mediating factors are inoperative. This is a warning to parents and other people charged with the care of children, rather than a reason to make television a scapegoat.

Yet even in the case of children, one of the earliest and most comprehensive investigations of the role of television in children's lives — Schramm, Lyle and Parker (1960) — concluded that 'it is the children who are the most active in this relationship. It is they who use television rather than television that uses them ... so when we talk about the effect of television, we are really talking about how children use television'. My own research findings, 23 years later, echo that view.

ADVERTISING: A TEST CASE

But what about advertising? At first glance, advertising looks like the most *deliberate* form of TV propaganda that, far more than any program, pushes the power of the medium to its limit. So how effective is TV advertising in shaping our tastes, preferences and values?

For a start, we need to acknowledge the sheer weight of TV advertising required to achieve any result. Sweeney Research (2001) estimates that children are exposed to about 350 000 TV commercials by the time they leave high school. Currently, on television alone, Australian advertisers spend almost $3 billion each year pleading with us to buy their wares. (Obviously, the television bug is not especially powerful, or it wouldn't need to be deployed so relentlessly and in such vast numbers.) In fact, TV advertising is such a *weak* source of direct influence on consumers that advertisers have to spend a king's ransom to achieve significant results. In 2000, two of Australia's biggest-spending advertisers, Telstra and Coles Myer, spent about $270 million between them, at a time when their shares of their respective markets were actually declining. You could argue that their market performance would have been worse if they had *not* advertised so heavily, but you certainly couldn't argue that all those millions of dollars had persuaded — let alone intimidated — the consumer into blind obedience to the messages coming at them with such frequency and force.

More than 80 per cent of new products fail. Some of these do not receive much advertising support, it's true, but even some of the largest and most sophisticated marketing organisations have found that new product launches are a risky enterprise with a high probability of failure. Clearly, when people say that 'advertising makes us buy things we don't really want', that proposition is not strongly supported by the facts.

So what are the facts? First, we need to remind ourselves, yet again, about that 'nexus of mediating factors'. What influences consumer behaviour most directly and powerfully is an impressive range of real-world factors — especially personal experience of the product — that do not have to be mediated through television or any other channel of indirect influence: force of habit, the pack, the price, promotional activity like discounts or other rewards and incentives, in-store displays and demonstrations, free samples in the letterbox, parents' example throughout childhood, friends' recommendations ... and, of course, the performance of the product itself. Advertising, by comparison, is a feeble influence, largely confined to reinforcement of existing attitudes and beliefs — shoring up the conclusions consumers have already come to — by preaching to the converted (always the most effective form of preaching).

Attempts to generalise about the effects of advertising have come to conclusions remarkably similar to Klapper's. In a seminal article entitled 'What do we know about how advertising works?' (included in a collection of significant papers published by the European Society for Opinion and Marketing Research), the eminent British market researcher, Timothy Joyce (1980), drew eight conclusions:

1 The view that 'advertising works by converting more people to use the brand' is misleading in several ways. Advertising may be working even though sales are level or even going down, by preventing (greater) loss of users. And, in some markets at least, increasing the loyalty of people who are already users, and increasing the amount they buy, may be a better prospect than bringing over non-users.

2 It therefore seems to be true that in many situations, advertising works by exploiting and reinforcing attitudes of people who may [already] be 'users' in at least a broad sense.

However, in other situations it obviously has an important role in extending trial.

3   Attitudes influence purchasing, but purchasing influences attitudes as well. Using research to establish the precise link between the two may therefore be difficult.

4   Consumers' decisions cannot be fitted to a model of rational choice, at least not without extensive modification of such a model. The 'rational argument' model of advertising is therefore generally inappropriate.

5   Attention and perception are highly selective. Consumers bring preconceptions to advertisements and may misperceive or misunderstand them.

6   Involvement in an advertisement is a much more complex matter than such terms as 'liking' and 'belief' imply: it probably does not matter if an advertisement is not liked or not literally believed. However, 'interest' in the sense of stimulation, and 'identification' in the very broadest sense, are probably important.

7   Recall of the product, rather than the advertisement or slogan, is what counts.

8   Above all, the consumer is not passive, helpless advertising fodder. There is a strong drive for consistency and stability in the consumer's structure of needs, attitudes and behaviour. This may lead to a tug-of-war between the perceptions of an advertisement and [existing] attitudes, and a tug-of-war between attitudes and purchasing behaviour.

Putting Klapper's and Joyce's conclusions together, we can begin to see how the most overtly persuasive media content — advertising — actually works.

The relationship between an advertisement and a consumer is complex and emotional, and it has more to do with consumers projecting their needs, attitudes, wants, values or aspirations on to an advertisement, than it has to do with an advertisement 'injecting' some want or aspiration into consumers.

Where consumers' existing attitudes and dispositions do not accord with the message of an advertisement, that message is unlikely to influence the consumer because it will be judged to be either irrelevant or 'false' (or, possibly, 'stupid').

Advertising appears to be most effective when it is designed to reinforce *existing* attitudes and values in the consumer — especially when it is designed to reinforce existing purchasing behaviour.

Since advertising represents the most extreme version of the television bug hypothesis, it warrants special attention; after all, the whole purpose of commercial television is to deliver an audience to the advertiser. That is what commercial television *is* — an advertising medium — so if anyone is going to know how to harness the power of the television bug, it will be the commercial operators. Certain conditions are imposed on licence-holders to ensure that there is some attempt at balance in their programming, but their primary goal is to broadcast programs that will maximise audiences for advertisements. The networks are not looking for the warm inner glow that comes from attracting big audiences: they are looking for revenue, and the prices they can charge for TV air-time are utterly determined by the quantity and quality of the audience they can deliver to the advertiser.

So let's examine a couple of cases where advertising appears to have had a dramatic effect on consumer behaviour — not in the transparent sense of retail advertising, where the purpose is simply to inform potential customers that a particular product is selling for a particular price, but in the more subtle area of so-called 'image' advertising. Consider the famous Meadow Lea case. (If you didn't know it was famous, consider this: in the late 1980s and early 1990s, the Meadow Lea TV advertising campaign was credited with increasing Meadow Lea's share of the margarine market from 11 per cent to 26 per cent — an almost unprecedented performance.)

How did they do it? The answer is simple: they strongly reinforced existing favourable dispositions towards Meadow Lea among occasional consumers of the brand who, in the process of having their favourable dispositions reinforced, gradually became more regular users. In other words, they preached to the converted. You can hardly imagine a more nurturing, reassuring, affirming message than their slogan: 'You ought to be congratulated'. (Some unkind souls have claimed that the slogan was a lucky accident, arising from its creators' desperate search for a word that rhymed with 'polyunsaturated').

Range Rover is a less obvious but equally impressive example of an advertising campaign that changed people's behaviour quite radically by reinforcing beliefs they already held. It's another case of preaching to the converted — in this case, not existing owners of the brand so much, but people whose existing dispositions made them responsive to the brand's message. Range Rover led a revolution that has seen Australians become the world's most enthusiastic buyers of four-wheel drive vehicles (most of which never leave the bitumen but still manage to imbue their drivers with an extraordinary sense of power and freedom). How did Range Rover do it? Mainly by producing an attractive and effective vehicle, of course, but advertising played an important role as well. The turning point was an advertising campaign based on the slogan: 'Write your own story'.

Contrary to conventional wisdom in the advertising industry, that was not a mind-*changing* campaign; it was a powerfully *reinforcing* campaign. It took an existing attitude among people who dreamed of owning a 'getaway' four-wheel drive vehicle and reinforced it. It fed their own dreams back to them: buy a Range Rover and you can indulge all your rural fantasies!

Perhaps the most counter-intuitive evidence about the power of advertising concerns the effect of alcohol advertising on consumption rates. As reported by Burrett (1990), research conducted by the Department of Medicine at the University of Western Australia shows that from 1976 to 1986, per capita consumption of alcohol by people over the age of 15 declined consistently (by about 10 per cent), even though advertising expenditure increased markedly (by more than 75 per cent) over the same period. (TV advertising, in particular, increased by 250 per cent in that period.)

Similarly challenging evidence comes from pre-unification Germany, where alcohol consumption in East Germany, with no advertising, was slightly higher than in West Germany, where heavy promotion of alcoholic beverages occurred.

The important underlying point here is that even when the media are at their most overtly persuasive, they are mainly reinforcing what people already think, feel or want. If there *is* a television bug, it's a bug which seems to lack the power to convert: its seductive power lies in its ability to encourage us to do what we want to do, be who we want to be, and dream what we want to dream. This is a highly indulgent, permissive bug.

Of course, you can raise serious ethical questions about whether some existing dispositions — for example, greed — *should* be reinforced, but it would be wrong to blame the media for putting those dispositions there in the first place. Similarly, the argument that advertising diminishes people's freedom by selectively reinforcing some of their attitudes and dispositions at the expense of others, overlooks the fact that consumers come to the marketplace with absolute sovereignty: they can turn their TV sets on or off, and they can embrace or ignore advertisers' messages at will. On the other hand, it's true that the purpose of most advertising is the exploitation of consumers for commercial purposes and, as Benn (1967) has argued, people are only 'free' if the persuasive messages they are exposed to do not involve 'hidden manipulation'. (If there *were* a television bug, then, you'd be morally obliged to warn the audience before you transmitted it!)

Of course, there was a time when people believed 'hidden manipulation' was perfectly possible, in the form of so-called 'subliminal advertising', in which words or images reach the consumer below the threshold of normal perception. This concept took hold of the public imagination (and, no doubt, of the imagination of a few rapacious advertisers as well) when it was alleged that the flashing of subliminal ice-cream advertisements onto the screen of a New

Jersey cinema in 1956 boosted ice cream sales at the interval. In spite of numerous attempts to replicate (or even to find evidence for) this experiment, it has finally emerged — as suspected by Dixon (1972) — that the whole thing was a hoax.

The hysteria surrounding the idea of subliminal advertising is a classic symptom of our lurking fear of the television bug, and of our scarcely suppressed media mania. It is almost as if we *want* to believe that manipulation is possible, even though the mounting evidence suggests that television's effects are largely benign.

When advertising isn't setting out to reinforce our existing behaviour, it is probably setting out to change our behaviour by reinforcing some existing attitudes and refocusing them on a different brand. That's what advertising does best ... but here's a more fundamental point: *that's the way the mass media work*. Program content tends to resonate with the audience's existing attitudes, values and pre-dispositions. The TV viewing experience has more to do with what the viewer brings to the program (or the ad) than what the program does to the viewer.

Before we leave the subject of advertising, consider the case of one of Australia's most famous political advertising campaigns — a campaign widely credited with having been so 'powerful' that it won a Federal election for the Labor Party in 1972. It certainly was an effective — even a powerful — campaign, but the power lay in its ability to reinforce what people were already thinking ('It's time for a change'), rather than changing their minds.

The 'It's time' campaign didn't *persuade* the electorate it was time to change; it simply asserted what most voters had already come to believe, and encouraged them to act on that belief.

Tony Schwartz (1973), a US advertising expert whose political commercials were said to have been influential in the re-election of President Lyndon Johnson, describes the relationship between television and viewers like this:

> A listener or viewer brings far more information to the communication event than a communicator can put into his program, commercial or message. The communicator's problem, then, is

not to get stimuli across, or even to package his stimuli so that they can be understood and absorbed.

Rather, he must deeply understand the kinds of information and experiences stored in his audience, the patterning of this information, and the interactive resonance process whereby stimuli evoke this stored information.

That sounds like the kind of thing Josef Goebbels might have said. His propaganda machine served Adolf Hitler's Nazi Party brilliantly, even though most observers from outside Germany thought Hitler a rather ridiculous figure and couldn't work out why he commanded such respect — such adoration — from the German people. Carl Jung was there at the time, and he knew why. This is what he said in an interview with the US journalist HR Knickerbocker (1939):

> It is because Hitler is the mirror of every German's unconscious but of course he mirrors nothing from a non-German. He is the loudspeaker which magnifies the inaudible whispers of the German soul until they can be heard by the German's unconscious ear.
>
> He is the first man to tell every German what he has been thinking and feeling all along in his unconscious about German fate …

A claim frequently made about modern society is that the media push particular agendas and move the community to embrace those agendas. While it is obviously true that media content fuels — and sometimes even structures — our conversations, it's hard to support the view that the media 'tell us what to think', in significant or enduring ways. At every election, for example, newspaper editorials — to say nothing of TV campaign commercials — thunder their messages in favour of one party or the other, but does this produce some kind of 'brainwashing'? In 1993, virtually every newspaper editorial in the country urged the election of the Hewson-led

Coalition and the defeat of the Keating-led Labor Government (and most commentators predicted that result), yet Labor won.

While the mass media may well play a role in promoting collective action (such as protests and demonstrations) by presenting examples of it to a wide audience, it would be naïve to assume that the media plant the idea of collective action in otherwise apathetic or neutral minds — let alone minds positively hostile to the message. When people see riots and sit-ins on television, they may be encouraged to copy those strategies for expressing their own outrage, but the outrage has to be there to begin with.

LET'S KNOCK TELEVISION OFF ITS UNWARRANTED PEDESTAL

I am no apologist for the mass media. But I am strongly resistant to the idea that the media should be a scapegoat for things that displease us about modern society.

The dollar's falling? Blame the media! The crime rate's rising? Blame the media! (Don't the blame the media when the crime rate's falling, though.) Children's manners are worse than they used to be? Blame the media!

The research evidence, over a sustained period of very fruitful activity, should lead us to be more cautious in our tendency to blame the media — especially television — for having a direct effect on the attitudes and behaviour of the community. Klapper's 'nexus of mediating factors' is real; in contemporary Australia, it includes things like the increasingly unequal distribution of work and wealth, social upheaval in the wake of economic restructuring, the far-reaching impacts of the gender and IT revolutions ... to say nothing of the destabilising effects of a soaring divorce rate, the lowest marriage rate for 100 years, the lowest birthrate ever, the shrinking of the Australian household, our epidemic of depression and a widespread sense of disconnection and alienation.

We can't blame the media for any of that, but we do need to remember that this is the climate, the environment, in which many people come to their TV sets looking for relief and comfort — the relief of a good laugh, the comfort of ritual, even the comforting reinforcement of prejudice — to say nothing of simple distraction.

We must resist a naïve view of our relationship with the media, especially one that casts us in the role of passive victims. We must acknowledge that in the relationship between television and you, the most powerful factor is *you*. When you view television, you *use* it; the way you view it — the way you use it — will obviously determine its role in your life. You may choose to use it as a nightly soporific, as growing numbers of people do. You may use it as a form of stimulation. You may use it as audio-visual wallpaper, creating a comforting background flicker of sights and sounds that reassure you by their very familiarity and predictability.

The point is that unless we have allowed ourselves to become helplessly addicted to it, we are in charge of television. We are viewers, not victims. And, according to recent research, most of us are positively ruthless in our role as *active* viewers: David Dale (2001) has reported some analysis by Sydney researcher David Keig, showing that, on average, only 36 per cent of those who dip into a program end up watching the whole show. Typically, 51 per cent of the total audience for any one program watches less than one third of it. Some programs attract more loyal viewers than others, but those who fear the invasion of the television bug might draw some comfort from the fact that viewers appear to be remarkably fickle and transient in their encounters with TV programs.

All this flick, flick, flicking suggests something less than close attention; it certainly suggests a model of TV viewing that challenges the viewer-as-sponge model. (I suspect there may be a gender difference here: give a man a remote-control wand and he can't resist channel surfing; give it to a woman, and she'll think it's merely designed to save her the trouble of getting up when she *needs* to change channels — perhaps on the disturbingly rational grounds that the program she was watching has ended.)

There's something faintly pathetic, shameful and absurd about our concern with the 'impact' of TV programs, as though we are not agents in the viewing process. When a TV program does have an 'effect' on us, this will usually be because we sought it out as a match for our existing mood, or as a mirror to our existing attitudes and predispositions, or as fuel for our prejudices, or as a comforting ritual, or as a vehicle for meeting our needs in some other way. Most of us, most of the time, use television to help us stay the way we are; to comfort us; to reassure us ... and, of course, to amuse and distract us. Under those conditions, television is rarely an agent of conversion, let alone mass brainwashing.

This is even true for children: in most cases, parents, teachers and peers are a more powerful influence on children than television is. Children may copy what they see and hear on the screen, but they quickly learn to distinguish between that and the three-dimensional reality of the world they live in. In most families, children who tried to act out the role of a TV hero or villain in an inappropriate way would soon be put in their place. Socialisation is a powerful converter; mass communication isn't ... another example of Klapper's 'nexus of mediating factors' at work.

We are gradually learning how to put television in its place — not only in our use of the medium, but in our thinking about it. We need to take television off the pedestal where we have placed it, and to acknowledge that the hopes and fears we had for the medium were probably unrealistic. Over time, most people come to regard television as inhabiting its own peculiar, idiosyncratic reality — 'another world', which some viewers enter willingly and habitually, and which others approach with a sense of caution, scepticism and reserve.

Whatever role you allow television to play in your life, the critical thing to remember — and to teach your children — is that *it's only television*. We should neither idolise not demonise it: both responses allow us to abdicate our own responsibility to use the media wisely and to guide and encourage our children towards wise use. The media will continue to do their job of trying to maximise their audience in any way they can; we must continue doing ours, as sceptical and discriminating viewers, as well.

The 1986 Mackay Report, *Television*, suggested that the process was well under way:

> Australian viewers have entered a new phase in their long-term relationship with television. Television is still integral with family life; it is still one of the most effective relaxants and escape-hatches from the demands of daily life ... but television has lost its edge of excitement. Television has settled into our lives. Memories of the impact it made on us in the late-fifties and early-sixties are now fading; we are less involved in the idea of television, less committed to it, more inclined to accept it as a permanent member of the family, with all its well-known strengths *and* weaknesses ... We no longer have great expectations of how the medium will change our lives or 'improve us'. We use it in a less ambitious and less optimistic way ...
>
> Contemporary television is cheerfully tolerated rather than loved; accepted rather than admired; appreciated rather than embraced.

Five years later, our report on *Our Evolving Relationship with Television* noted that 'Australians' evolving relationship with their TV sets is like so many personal relationships: they mature; passions cool; adjustments are made; expectations are lowered; we get on with it'.

Still, the fact that television is not 'powerful' in the way it has often been presumed to be should not blind us to the fact that addiction to media content is as dangerous as addiction to anything else, because it saps our spirit and weakens our resolve. We need to guard against easy acceptance of the idea that, when nothing else is happening, 'there's always something on television' — regardless of what it is. The real issue of the media in society is not so much about what's going on in the media: it's about what's going on in society and, in particular, what's going on in the lives of individual viewers. The state of our lives determines what we bring to the encounter with television and, in turn, what we take away from the encounter.

Perhaps we should follow the example of today's teenagers who treat television rather as they treat the refrigerator. Like the fridge, they leave the TV set running continuously without paying much attention to it most of the time; like the fridge, they take out of it what they want, when they want it.

2

WHAT KIND OF VIOLENCE?

The world is a less violent place than it used to be. Rape and pillage still occur but, war zones aside, the human race seems gradually to be becoming more civilised. I wouldn't go so far as to suggest that the modern phenomenon of media violence is the reason for this, and I am sure you would be outraged if I did. But it's precisely that kind of coincidence — more violence on television, less in the real world — that is easily confused with causation.

Strangely enough, conventional wisdom runs in the opposite direction on this particular subject — especially when it comes to children. There's a widespread belief that children are becoming increasingly aggressive and that their aggression is being fuelled by a steady diet of media violence. Everyone can give you an example of a particular child or adolescent who has become violent *because of* its exposure to certain TV programs, movies, or video games. (If they can't give you a particular example, they will confidently assert that this is known to be a *general* problem.)

How easily we gloss over the alternative possibilities! We seem almost intuitively to go for the causal explanation first. But isn't it possible that children with a violent disposition — created by all kinds of psycho-social factors — might have an insatiable appetite for media violence? Isn't it possible that the causes of real-life violence might be more closely associated with mental illness, family dysfunction, poverty or drug addiction than with media consumption?

What will we say about the systematic abuse suffered by some indigenous women and children, as reported by Rosemary Neill (2001)? A Queensland Government report, *The Aboriginal & Torres Strait Islander Women's Task Force on Violence*, published in 1999, concludes that 'increasing injuries and fatalities as a result of interpersonal violence … threaten the continued existence of Australia's indigenous peoples'. You can hardly imagine a bigger claim than that, and I am sure you would flinch from any simplistic explanation: how would you react if I said that a significant — or even the major — cause of violence in Aboriginal communities was exposure to violence on television? Would you be incredulous?

And what about the findings of the new US National Institute of Health study, reported by Adele Horin (2001), which seem to blame childcare for childhood violence? This study suggests children who experience long hours of childcare were more at risk, by kindergarten age, of developing behaviour problems such as aggression and disobedience. Not only were these children apparently more likely to be assertive and defiant, they were also more likely to bully, fight with and 'act mean' toward other children.

No doubt you imagine there is more to this than meets the eye, since so many children emerge from the childcare experience without such behavioural problems. And you'd be right to be sceptical: there are usually several factors involved in the explanation of any piece of human behaviour, which is why it is simplistic and unrealistic to try to isolate television — or any other single factor — as the sole cause of some behavioural problems in children.

While we're considering the range of alternative explanations for the presence of violence in the community, we should also register the unfashionable fact that aggression has a legitimate place in the spectrum of human emotions. It's built in and, under certain conditions, it comes out. What is not entirely clear — yet — is whether to let it out vicariously, via the consumption of media violence or the playing of mock-violent games by children, might actually *reduce* the risk of an eruption of real-world violence. Who knows whether exposure to media violence might actually be cathartic for many children? Most children who watch violence on television *don't* behave violently: should we interpret that as a good sign? (If so, a good sign of what? A sign that television is doing its cathartic work, or that children's personal and familial values are stronger than those they see portrayed on the screen?)

In any case, the proposition that media violence has made us a more violent society depends upon our belief that society *is* more violent than it used to be. In Australia, the current homicide rate is slightly over half the rate of 100 years ago and the NSW Bureau of Crime Statistics and Research frequently finds itself having to defuse public hysteria about the rising crime rate: some kinds of crime are up, some are down, and the explanation for the current increase in crimes of personal assault seems to have more to do with thefts related to drug-addiction, or the aimlessness of unemployed and alienated young people, than with exposure to media violence.

In the United States — widely regarded as a violent society helped along by an excess of violence on television — the rate of violent crime has been in its most sustained decline since World War II (LaFree, 1999). The marked drop in the crime rate during the 1990s is particularly surprising because, to quote Gary LaFree, 'it comes on the heels of dire predictions about the rise of a generation of "superpredators" who would soon unleash the full force of their destructive capacities on an already crime-weary nation' (for example, Fox & Pierce, 1994).

In any case, if you're worried about having your bag snatched at the bus stop, imagine the hazards of life in some of the neighbourhoods that once caught the eye of Vlad the Impaler: throughout history, violent people have managed to devise terrible ways of venting their spleen without the media examples of Donald Duck, Jackie Chan or Godzilla to inspire them. Jack the Ripper was not a video junkie.

Yet the suggestion that TV violence directly causes violent behaviour, especially in children, is persistently made — most recently by the Australian Early Childhood Association in their 2000 publication, *The Effects of Media Violence on Young Children*. Although it correctly acknowledges the *primacy* of genetic inheritance and the family environment as influences on a child's behaviour, the report also contains some disturbingly naïve generalisations, presented as if they support claims for the negative effects of TV violence. These assertions attracted some media attention because they so closely coincide with many of our prejudices on this subject: 'Children play games in which they pretend to be [media] heroes', the report thundered, 'and somebody inevitably gets really hurt'. (Really? *Inevitably*? More than playing, say, netball?)

Violent program content on television obviously gives children a ready-made framework for violent play, but violent play is hardly anything new. (I recall participating in frequent cops-and-robbers activity in my own pre-TV childhood. We regularly plugged each other full of imaginary lead and cheerfully assumed the identities of goodies and baddies without appearing to sustain any lasting damage … but perhaps I missed something.)

Such assertions, though well meant, do not address the more basic issue: is violent play harmful or beneficial? Both possibilities are equally plausible, depending on the circumstances and psychological state of the children concerned. Some disturbed viewers, young or old, will have their predispositions toward violent behaviour reinforced by what they see on the screen, but such dispositions can find reinforcement from all kinds of other places, if television doesn't happen to be available.

There's one bit of statistical evidence that is often overlooked when people want to draw a direct line of influence from media violence to violent behaviour: the evidence that suggests the people most likely to be involved in violent behaviour — either as perpetrator or victim — are young males. It is young males, however, who are the *lightest* TV viewers in the community.

ACNielsen's 2000 analysis of TV viewing trends revealed that people in the 16–24 age group spent less time than any other age group watching television. The analysis showed that 16–24-year-olds (the group most at risk of involvement in violent crime) watched an average of 2 hours 15 minutes per day, compared with the average for all people of 3 hours and 13 minutes. (By the way, those figures can also be compared with the 12 minutes a day spent by the average couple talking to each other, according to the Sweeney Research report, *Eye on Australia*, 2001.)

By contrast, the people most heavily exposed to television (and therefore to TV violence, presumably) are those in the over-55 age group who, for reasons you can easily imagine, spend an average of 4 hours 18 minutes per day watching television. If you believe in a powerful, causal link between violence on television and violent behaviour, perhaps you should cross the street, or hide somewhere the next time you see a person over 55 approaching.

Before we go more deeply into the relationship between media violence and violent behaviour, it's worth noting which programs dominate people's viewing. In 2000, the top TV programs for children aged 5–12 were: 'Popstars', 'Sabrina, the Teenage Witch', 'Family Guy', 'Futurama', 'Treasure Island', 'The Simpsons', 'Groundforce' and 'Home and Away'.

In the 13–17 age-group, the top programs were: 'Friends', 'Popstars', 'Spin City', 'Dawson's Creek', 'Shipwrecked', 'The Simpsons', 'Treasure Island', 'Family Guy', 'Charmed' and 'The Mole'.

The more you look at the programs young people watch in large numbers, the less likely you are to feel concerned about the possibility that television is making us a more violent society. Those heavy viewers over the age of 55 are a bit more disposed towards violent material; their favourite programs in 2000 were: 'Monarch of the Glen', 'Burke's Backyard' and 'The Bill'. (Perhaps it's 'The Bill' that causes all the trouble.)

Messner (1986, 1987) and Blau (1987) have approached the relationship between TV violence and violent behaviour from an unusual perspective. They assert that crime is less frequent when the routine activities of potential offenders and their victims reduce their opportunities for contact. In other words, any activity (such as watching television) that separates those who are prone to violence from each other, or from potential victims, is likely to decrease the amount of violence in the community.

TV viewing is certainly one way of keeping people away from each other. Since people mainly watch television at home, the opportunities for violence — at least with people outside the family — are probably reduced. Even the opportunities for domestic violence may be reduced, since people watching television a great deal are probably interacting less often with other family members. Messner found that cities with high levels of TV viewing have lower rates of both violent and non-violent crime.

It is possible to pick a number of holes in that particular theory — especially because aggregate analyses of this type overlook the specific viewing habits of those involved in particular instances of violent crime. But the theory raises an important question: does viewing television *reduce* the risk of violent behaviour, even among people predisposed to behave violently?

That's a question worth pondering, especially when you consider some of the evidence that is often cited to suggest a strong causal link between media violence and violent behaviour.

ATTEMPTS TO PROVE THE LINK BETWEEN MEDIA VIOLENCE AND VIOLENT BEHAVIOUR

Arguably the most relentless feature of mass communication research over the past 30 years has been the persistent attempt to design experiments that will demonstrate a direct, causal link between violent program material and violent behaviour. Some of the experiments seem to me to be laughably naïve in their attempt to set up conditions in which violent behaviour will emerge in direct consequence of exposure to violent material — though it always depends on how you define violence.

One of the most famous, formative experiments (Bandura & McKenzie, 1963) involved two groups of children: one group was shown a film in which children were seen punching an inflatable Bobo doll; the other group was not shown the film. Each group was then taken into a room where Bobo dolls were present and — are you surprised by this? — the group that had seen children punching Bobo dolls in the film was more inclined to punch them than the children who had not been exposed to the film.

Personally, I wouldn't be inclined to make too much of that copycat effect, especially since it is debatable whether punching an inflatable Bobo doll is what we normally mean by violent behaviour (Tedeschi et al., 1974). But, believe it or not, that particular experiment has heavily influenced the direction of the debate about media violence over the past 30 years.

Other attempts to prove the link between media violence and violent behaviour have involved such things as photographing the expression on children's faces while they view violent material and then establishing, by subsequent observation, that those who looked happy or relaxed while being exposed to violence were more likely to behave in violent ways in subsequent social settings. Note the tendency here — which frequently occurs in experiments of this kind — to assume that it was the violent material that *caused* the subsequent violent behaviour, whereas, even in that bizarre experiment, you would have to admit the possibility at least that children who expressed pleasure while watching violence were more heavily predisposed to behave violently.

In general, laboratory experiments are unsatisfactory ways of examining the effects of media violence. Most studies conducted under laboratory conditions do show that subjects who observe media violence tend to behave more aggressively than the subjects in control groups. But the research is inconsistent in showing whether it is necessary to provoke subjects *before* showing violence to get an effect — in other words, it is not clear whether media exposure acts as an instigator of aggression in the laboratory, or merely as a facilitator. (Friedman, 1984)

And that's the problem: laboratory situations are significantly different from real-life situations outside the laboratory. Offering an opportunity for someone to behave violently in a laboratory setting might tell us something, but it is unlikely to tell us much about how they might behave in the real world. The famous — and frightening — experiments conducted by Stanley Millgram (1973) illustrate this difficulty: although subjects in Millgram's experiment seemed remarkably willing to administer an electric shock to victims viewed through a one-way screen, that may tell us more about the nature of laboratory experiments than about human behaviour in the real world. (It does suggest, however, that we are all too ready to respond to instructions given to us by people in positions of authority — whether Nazis in the Third Reich or experimenters in white coats.)

A number of more imaginative experiments have been conducted outside the laboratory — though, of course, they have to use subjects whose behaviour can be carefully monitored and this often confines the studies to institutional settings. For example, studies have been conducted (Feshbach & Singer 1971, Leyens et al. 1975, Parke et al. 1977) purporting to show a causal relationship between boys' exposure to violent or non-violent programming and their subsequent levels of aggression. But such studies have important methodological limitations (Freedman, 1984) because they assume that each subject's behaviour is independent, whereas, in fact, they are living and interacting together in ways that may well exert more powerful influences on their behaviour than the influence of television.

Such studies tend, in any case, to be very inconsistent in their findings. Yes, you can find studies that show a positive correlation between exposure to violent program material and an increase in violent behaviour, but you can also find studies that show the opposite: exposure to violent TV programs being associated with *less* aggressive behaviour (Feshbach & Singer, 1971).

Now that so much of this kind of research has been undertaken, meta-analysis has become possible. Wood (1991), for example, analysed 28 studies where children or adolescents had been observed unobtrusively after being exposed to aggressive or non-aggressive film material. In 16 of these studies subjects appeared to engage in more aggressive behaviour following exposure to violent films, while in seven studies subjects who were *not* exposed to the violent material engaged in more aggression, and in five studies there was no difference between control and experimental groups.

Some of the most imaginative studies of this relationship have tried to establish the impact of the arrival of television on levels of violent behaviour in full-scale, real-life communities. (Note that all the studies assumed that high exposure to television involves high exposure to TV violence — again, there is the problem of subject assessment and interpretation, and the ever-present risk of over-simplification: is 'The Simpsons' a violent program? Is 'Bugs Bunny'?)

A 1982 study (Hennigan et al.) compared crime rates in American cities that had television with those that did not. No effect of the presence or absence of television was found on violent crime rates in a comparison of the two kinds of cities. Furthermore, when cities without television obtained it, there was no increase in violent crime. There was, however, an increase in the incidence of theft, which the authors attributed to the feelings of relative deprivation suffered by viewers observing affluent people on television.

But other studies have been less conclusive than that one. An intensive study (Joy et al., 1986) was undertaken of the behaviour of children after television was introduced to an isolated Canadian town in the 1970s. The town was compared to two others that were judged to be similar, but already had television. About 45 children in the three towns were observed on the school playground in first and second grade, and then again two years later. The researchers reported that the frequency of both verbal and physical aggression increased in all three communities, but the increase was significantly greater in the community in which television was introduced during the study. This was a complex piece of research and some of the results reported were not consistent with the TV effect: for example, in the first phase of the study, the children in the community without television were just as aggressive as the children in the communities that already had television, so the picture was confused from the outset.

Yet another line of investigation is to try to establish the impact on violent behaviour of highly publicised violent events reported in the media. There is some evidence to suggest a copycat effect in everything ranging from violence in response to the telecast of a prize fight (Phillips, 1983) to anecdotal evidence (or mere rumour?) purporting to suggest an increase in youth suicide attempts when a high profile pop star dies by his or her own hand.

But, almost without exception, controversy rages about the methods used to measure such phenomena. The best we can say is that there is no automatic or general effect, but there are signs of occasional effects, specific to particular events in particular circumstances. (Refer again to Joseph Klapper's five generalisations in Chapter 1.)

The Canadian researcher Jonathan Freedman (1984), among others, has found a positive correlation between the amount of exposure to TV violence and the frequency of aggressive behaviour, but, to quote from Richard Felson's (1996) analysis of this work:

> There are good reasons to think the relationship is at least partly spurious. For example, children with favourable attitudes towards violence may be more likely to engage in violence and also more likely to find violence entertaining to watch. Also, children who are more closely supervised may be less likely to engage in violence and less likely to watch television. Intelligence, need for excitement, level of fear, and commitment to school are other possible confounding variables.

But even when strong positive correlations can be found between the viewing of media violence and aggressive behaviour, there are almost always other factors to take into account. Intelligence is evidently one of them: some researchers (Wiegman et al., 1992) have found low intelligence to be associated both with the level of consumption of media violence and the incidence of aggressive behaviour — that is, less intelligent people tend to watch more violence *and* to behave more aggressively. But does one cause the other, or is low intelligence a causal factor in both situations, independently of each other?

WHEN THE EXPERTS DISAGREE, IT'S WISE TO REMAIN SCEPTICAL

The confusions and contradictions in the data should not lead us to conclude that there is *never* a causal link between media violence and violent behaviour, but it should encourage us to be extremely cautious about any of the assertions we are inclined to make. Violent behaviour is the result of a very complex array of factors: it is true that some violent people seem to have a voracious appetite for violent media content, but many non-violent people also enjoy media depictions of violence. Not everyone who watches violence will behave violently, any more than people who watch comedy will become funny, people who watch voyeuristic programs will become voyeurs, enthusiastic viewers of cooking programs will become *cordon bleu* chefs, or insatiable viewers of football will rush out and become footballers.

It's more complicated than that, in every case, and we do ourselves an intellectual disservice when we assume that the effect of violence in the media is a special case, uniquely able to be explained in simple terms.

This is not to dismiss the possibility of some causal links between media violence and real-world aggression, but we would do well to be sceptical and, in particular, to note that most of the studies that try to prove *causation* end up proving no more than *correlation*.

One of the staunchest advocates of the causation line, Albert Bandura (1983) — the man who gave us the Bobo doll experiment — argues that TV content can determine the forms aggressive behaviour will take. The copycat effect cannot be denied but, again, it would be naïve to assume that a person disposed to behave violently only did so because television had provided a blueprint of how to go about it. The history of the human race, pre-television, rather argues against this thesis.

The question of whether exposure to TV violence produces, over time, a *desensitisation* to violence is also controversial. It appears to be true (Thomas et al., 1977) that heavy viewers of TV violence respond less emotionally to violence than light viewers do: does this mean that they are simply bored with it, because they've seen so much of it, or is there some emotional desensitisation going on? It's possible that desensitisation might weaken the effect of a heavy diet of TV violence — the more you watch, the less it means to you — but this is speculative and, in any case, it depends on the assumption that viewers cannot tell the difference between media violence and the real thing.

In reviewing the range of available research into the relationship between mass media and violent behaviour, Felson concludes that 'the inconsistencies of the findings make it difficult to draw firm conclusions about the effects of exposure to media violence on aggressive behaviour'. He acknowledges that most scholars who have reviewed research in the area believe there *is* an effect, but many other scholars have concluded that the effects are purely correlative and not causal.

Still, as Felson acknowledges, it would be surprising if pervasive media violence had *no* effect on viewers. In terms of Klapper's original conclusion about the capacity of the mass media to reinforce existing dispositions, it would be perfectly reasonable to argue that TV depictions of violence might well reinforce violent predispositions among some viewers: some people, because of what they bring to the experience of viewing media violence, are more likely to be affected by it — directly or indirectly — than other people. Sitting in your dressing gown and slippers in front of 'The Bill' on a Saturday night, sipping cocoa, is unlikely to be a violence-arousing experience for older viewers. On the other hand, aggressive young males who are looking for outlets for their aggression might, under certain circumstances, copy aspects of things they have seen in the media to structure — or even to embellish — their own aggressive behaviour.

Among children, violent *play* is almost universal and, throughout history, children have borrowed from the adult world — more recently, via the mass media — to give structure and content to their play. Once it was King Arthur legends, then it was cowboy comics, then it was radio serials and now it is television. The language and gesture of play might change according to cultural shifts and fashions; the underlying tendency to play violent and/or aggressive games seems to be persistent.

Felson's comprehensive analysis of the literature of media research leads him to conclude that 'the failure to find individual-difference factors that condition the effects of media exposure on aggressive behaviour contributes to scepticism about media effects'. Scepticism is clearly in order.

In the light of the available research we can say that there are some copycat effects of media content — including media violence — and that most of these occur, at least in children, in the form of play. We can speculate about the relative incidence of media violence providing a channel for the harmless discharge of violent emotions among adolescent and adult viewers: media violence may well play an important surrogate role in our society. We can even speculate about whether exposure to media violence increases aversion to violence, or desensitises us to it: both propositions are plausible; neither is proven.

Without wanting to downplay the issue of media violence, or the importance of monitoring its effects, I sometimes wonder whether this debate is as much about aesthetics as morality. We are entitled to like or dislike any type of media content, and we are entitled to encourage our children to watch some types of material and avoid others. (Personally, I detest on-screen violence, and I hate the real thing even more.) But that is different from claiming that program content we dislike is damaging.

It goes without saying that if you find yourself disturbed or distressed by any program — violent or otherwise — you should turn it off. If you see signs of your children being upset by violent material, you should offer them a more palatable alternative. (I tend to avoid garlic and chilli in food as well.)

On the other hand, if your kids are racing around pretending to shoot each other, or copying other acts of TV violence, perhaps you should be thankful their aggression is being discharged so playfully.

If we really believed that media violence *causes* aggressive behaviour — in a direct and demonstrable way — why wouldn't we ban it? Perhaps it's because we have noticed that the vast, overwhelming majority of TV viewers, whether of violence or anything else, are law-abiding citizens. Perhaps we've also noticed that vast numbers of non-violent TV viewers, for whatever reason, actually enjoy watching violence.

Why that insatiable appetite for media violence exists may be connected to our more general appetite for information about disasters — rapes, murders, earthquakes, plane crashes — that allow us to catch a glimpse of the extremes of human experience, especially the dark side. What appalling things are humans capable of *doing*? What must it feel like to be the victim or the perpetrator of violence? These are legitimate questions to ask, and many people are curious about the answers. Perhaps media violence, whether documentary or fictional, satisfies an important, and generally harmless, thirst for more knowledge and more understanding of what is fully meant by 'the human condition'.

IS THERE ANOTHER *KIND* OF MEDIA VIOLENCE?

In his book *The Moral Sense*, Harvard philosopher James Wilson (1993) says this:

> The real problem with prolonged television viewing is the same as the problem with any form of human isolation: it cuts the person off from those social relationships on which our moral nature in large part depends.

Recent research conducted by Dr Tom Robinson and a team from Stanford University (reported by Baker, 2001) suggests that some of the adverse effects attributed to children's exposure to the electronic media — effects such as violence, teasing and bullying — could be reversed solely by decreasing the amount of exposure to television, videos and electronic games. Dr Robinson's research compared 105 children in one school who reduced their exposure to media material for six months with 120 children in another school who did not. (Both groups were from similar families in San Jose, California, and all the children were aged around seven and eight.) The group whose media exposure was cut exhibited better behaviour in playground and other social activity. Members of the reduced-exposure group were involved in about half as many incidents of aggressive playground behaviour as the second group.

At one level, this looks like a contradiction of Messner and Blau's assertion that TV viewing is likely to reduce the incidence of violent behaviour, by keeping people who are predisposed to violence apart. One key difference is in the age groups involved: Messner and Blau were concerned with adolescents and young adults, and Robinson's focus was exclusively on 7-and-8-year-olds. In both cases, it is important to note that the issue seems to be *TV viewing*, per se, rather than the viewing of violent material (though most researchers in the field would argue that heavy exposure to television effectively means heavy exposure to TV violence).

What the Stanford study is really saying is that if children spend more time with each other, and less time with television, they will learn to get along better with each other.

Why would this surprise us? After all, as James Wilson has suggested, the biggest problem about media consumption is that it takes us away from each other: when young children spend less time socialising, they are missing out on important learning about the process of socialisation, and so they are less likely to be good at managing conflict, negotiation, and all the other arts of personal relationships.

Wilson's point is simple and obvious, yet profound: the time we spend interacting with media content of any kind is time we are not spending with each other. If we know one thing about this whole subject of media power, we know that people exert more influence on each other than the media do. When time spent with the media becomes excessive — especially in our formative years — it poses a threat to the quality of our personal connections with each other. Thus, although some researchers have found that the more time people *with violent dispositions* spend with television, the less time they have to bash each other up, the more general point — for the majority of people lacking violent dispositions — is that the amount of time we spend interacting with the mass media is likely to have a direct, negative bearing on the quality of our relationships with each other. If we spend too long sitting in front of the box at the expense of time we might spend talking and listening to each other, our relationships are likely to suffer. Relationships take time to nurture, and television is a notorious thief of that time. In this sense, the media appear to be capable of inflicting a more subtle kind of violence on us than the kind we often fear.

Are we spending less time with our children because we — and they — are addicted to television and computer screens? Are we spending less time with our spouses for the same reason? Are we visiting our neighbours less, engaging in fewer community activities, and generally feeling less 'in touch' with each other because we are falling for the illusion of interconnectedness via the worldwide network of electronic data?

There seems little doubt that this form of media violence has the potential to erode the quality of our personal lives, and that has nothing whatever to do with the content of the programs we watch. There's a deeper question here: why do so many people seem to *prefer* to interact with the mass media than with each other? Is it simply because media babble is so relentless that we feel somehow obliged to give it our attention? (After all, a real person can be put on 'hold', whereas media content is essentially fleeting: 'I've got to watch this now, because it's on now'.) Or is it perhaps because in our personal relationships, we are now obliged to compete with mass-media content: do we have to appear to our children to be more amusing, creative, imaginative and entertaining than the things they watch on television? (No contest at my place, I'm afraid.)

Or is there something about the media — especially television — that seduces us with the idea that, at any moment, we are going to see something so significant or interesting that we should stick around? This may be one of the most violent effects of the mass media on our attitudes and values: the creation of the 'entertain me!' syndrome.

Perhaps we have been looking in the wrong place for the evidence of the effects of media violence, or perhaps we've been thinking of the wrong *kind* of violence. I suspect our relationships are being violated by the hidden message — the subtext — inherent in the mass media, especially television: the message that *everything should be entertaining*. The idea that education, politics, religion or *anything* should be made entertaining in order to be palatable to the mass audience represents a media effect far greater than the portrayal of a bit of physical violence in a crime drama.

This is the great seduction of a modern, media-saturated society. It is a dangerous tendency when, as a result of over-stimulation through the mass media, we begin to feel some disappointment in lower levels of stimulation, as though something is missing if we're not being amused, shocked or otherwise entertained. If the media give us a particular kind of buzz, then it's easy to see how, if we spend too long in front of the television, we could attune ourselves to the need for *constant* buzz. (This is precisely why people become addicted to the mental states induced by hallucinogenic drugs, powerfully amplified rock music and other mind-altering circumstances: 'coming down' is no fun at all.)

'*Boring!*' has become the mindless, reflex reaction of those who are disconnected, however fleetingly, from the entertainment machines that sustain them. The TV set, humming incessantly, feeds the illusion that something is always happening: there is no down-time; no time to think, to reflect, to ponder.

If all the world's a stage, many of us appear to have opted for a seat in the stalls: we have become obsessed with the idea that everything — politics, religion, sport, education, sex — should entertain us, and this is perhaps the greatest of the various forms of media violence. It has created our insatiable appetite for celebrities to glow and glitter in the media spotlight for a few minutes, until someone more entertaining comes along, and it has conditioned us to the idea that *pace* is an inherent virtue: the short grab, the quick cut, the fragment of an image that does no more than suggest the whole. Isn't the trivialisation of issues itself a form of cultural violence?

Electronic media saturation has shortened our attention span: the five-second grab is no joke. I was recently interviewed for a TV current affairs program — taped, unfortunately, rather than live — and after I had given what I thought was a tight, taut, disciplined answer to one of the journalist's questions, she said: 'I'll just ask you that again, and could you give us a shorter answer?'. What's really being called for, of course, is 'yes' or 'no' ... or perhaps just a grunt or subtle tilt of the head that does away with words entirely.

Because we are constantly rushing on to the next story — *what's next? what's next?* — our folk memory is in danger of being violated as well. Our preoccupation with the new runs the risk of making us impatient with the old. Innovation becomes inherently attractive and the things we might once have thought of as having enduring value may come to seem tired, slow and lacking in headline-grabbing appeal.

The 'entertain me!' syndrome violates our values. The more stories we hear, the less each one matters and the more we run the risk of reducing our capacity to interpret and judge. Our moral clarity is dulled by the sheer volume of media messages to which we are exposed. The quantity of media content becomes a distraction from its quality.

So here's another way of thinking about media violence: it is shortening our attention span by offering us summaries of every situation, every issue, which are so short that, over time, we come to expect that every issue *can* be captured in a brief sound bite or a fleeting visual image.

This form of media violence — quite unrelated to the content of specific programs — even attacks our vocabulary, by showing us how to compress everything into the fewest possible number of words, by accommodating the vocabulary of the youngest and least educated members of the audience and by replacing, as far as possible, words with pictures: no story is deemed worthy of prominence in the TV evening news unless there are pictures to accompany it.

Media violence also creeps up on us in its constant search for an adversarial approach, especially in the coverage of current affairs. What current affairs 'story' (a revealing word in itself) would ever be regarded as 'good television' if it only presented one person's views or — horrors! — if two people who were expected to disagree turned out either to agree or to be respectful of each other's opinions. Drama, whether on stage or screen, has always relied on conflict to produce dramatic tension. The difference today is that virtually *all* TV programming, including current affairs, relies on conflict to create the kind of tension that will attract and hold viewers — even when the subject under discussion is the quest for reconciliation between indigenous and other Australians. This represents an insidious reinforcement of the idea that 'we need a bit of conflict to make things interesting' or even that 'the normal way to solve problems is to argue'.

And what about the subtle effect of an implied message that, at any hour of the day or night, entertainment, amusement and distraction are available to us?

The US critic, Sven Birkerts (1994), has described as his 'cold fear' the possibility that 'we are, as a culture, as a species, becoming shallow; that we have turned from depth — from the Judao-Christian premise of unfathomable mystery — and are adapting ourselves to the ersatz security of a vast lateral connectedness'.

To conclude this chapter, I want to propose three practical responses to the problem of media violence — however it may be interpreted.

First, don't worry too much about young people's exposure to explicit portrayals of violence in media content: the jury is still out on its likely effect, and the critical factor is not what the program does to the child, but what the child brings to the program. If you're worried about a child's behaviour, look first to its social and emotional context, rather than the TV programs it watches.

Second, we ought to be more concerned about the general problem of the *quantity* of time spent by children — or any of us — with the media, than with the specific question of 'what's on?'. There is more to fear from the media's power as a *replacement* activity than from its content, per se.

Third, we need to be on guard against the violence done to us by the media's seductive argument that for anything to be worthwhile, it must be entertaining. It would be a tragedy if we were to fall for that one.

3

'I'LL JUST CHECK MY EMAIL ... AGAIN'

Imagine a society without 'the media' — no mechanical or electronic means of sending or receiving messages — no print, no radio, no television, no cinema, no telephone and no Internet.

You don't have to go far back into Western history to find pre-media societies: even though the printing press was invented in the 16th century, mass literacy was a long time coming. For most of our forebears, speech and gesture were the only tools available for communication, and the communication culture was shaped by the dynamics of very local neighbourhoods — villages — in which direct personal contact was possible.

(For another perspective on pre-media cultures, you could consult anthropologists' accounts of the communication climate of tribal peoples living in primitive circumstances. Until quite recently, the Australian Aborigines were such a group.)

In a pre-media culture, people communicate through words, tone of voice, vocal pitch, intensity, rate of speech, but even more significantly through the nuances of facial expression, posture and all the subtleties of what we now call 'body language'. Art and music always play an important role in shaping any communication culture, but in pre-media societies, communication is intensely and essentially personal. Then, as now, 'a picture is worth a thousand words' (because what we see tends to engage our emotions more immediately than what we hear) but the 'pictures' are to be found in the visual qualities and characteristics of the people who are interacting with each other.

In a pre-media culture, we don't try to separate the person from the message. We interpret what is said in the light of who said it and how it was said. We recognise that 'meaning' is not in the words, but in the people who are using the words; the 'meaning' of a conversation grows out of the encounter and is embroidered by every aspect — sight, sound, time, location, ambience, smell, context — of that encounter.

Communication is therefore more emotional and less rational than in a culture dominated by the printed word. History and myth are blurred. Facts and feelings are not distinguished as though one is 'true' and the other is not. Stories may emerge from historical reality and, embellished in the re-telling, become part of legend and culture.

This does not mean that rationality is absent from a pre-literate culture, but it is not worshipped. When messages are not written down, they remain part of us — living and changing with us. Subjectivity and objectivity are in constant interplay.

Communication in a village or tribal culture has a more public character than in a print-based culture. Information is shared; knowledge is passed around, because sharing is integral with knowing: shared knowledge is as much a part of the communal experience as shared land. The concepts of individuality and privacy scarcely exist, at least in the terms in which Westerners understand them today. The culture is dominated by the idea of belonging to the group, the community, the tribe: the sense of personal identity is grounded in the group rather than the individual. In such a culture, communication is inseparable from the dynamics of personal relationships; it isn't regarded as a process of transmitting and receiving a disembodied thing called 'information'.

With the invention of movable type (Gutenberg in Germany and Caxton in England) and the advent of the printing press, new technology began to change this culture, in the way that new technology usually does. During the next 500 years, we gradually adapted to typography and embraced the rich possibilities offered by mass literacy — a phenomenon that only came to full flower in Western societies in the 20th century, courtesy of compulsory education. As we came to accept the printed word as the basis for mass education and mass communication, we underwent a profound change in our communication style, our communication culture and, in turn, our habits of thought. For a start, mastery of the relentlessly rational system of reading and writing requires a particular kind of thinking process — quite different from the creative, intuitive synthesis of aural and visual impressions normally involved in making sense of the world — and not everyone finds that easy. Thanks to the emphasis on the printed form, grammar and syntax became part of the process of learning the language (and 'grammatical errors', unimportant in a pre-literate culture, became an educational obsession).

As the village culture evolved into a print culture, we learned to separate the person from the message. Then, as mass communication became less personalised, we began to look for 'meaning' in the words themselves — sitting there on the page in black and white — not in the person who wrote the words. We submitted to the disciplines of the formal, linear, structured, logical style of the print medium — one word after the other, left to right, each line following the one before — and in the process, we tended to become linear thinkers. We were gradually converted into *homo typographica* — people who admire objectivity and enshrine rationality; people who say 'I'll believe it when I see it in black and white'; people who have to learn to 'read between the lines'.

This was both good and bad. On the one hand, typographic man — a description coined by Marshall McLuhan (1962) — became a mass phenomenon, conditioned by the medium to think in a linear, logical way, eventually creating a market for the one-man crusade by Edward de Bono (1971) on behalf of *lateral* thinking

— that is, the way we *used to* think, pre-print. On the other hand, widespread use of the print medium allowed messages to survive, intact, over space and time. The political, military, diplomatic, cultural and archival implications of that are self-evident. Historical records can now survive, intact, but, as Jonathan Miller put it in *McLuhan* (1971), 'the imaginative integrity of the sacred past no longer exerts such a comprehensive hold upon the individual members of a literate community'.

The literacy revolution fuelled by typography conditioned us to think of the 'best' or 'highest' form of communication as being a private, solitary, individual activity. Even if many people read the same thing, they read it individually and make their own private sense of it. When we feed on a steady diet of print, the relationship is between the ink on the page and the reader; the page actually becomes a mechanical barrier between writer and reader; each relates to the page, rather than to each other.

As we became readers and writers, we became more intensely individual; as we became more conscious of our individuality, privacy became a central social concern, and a cultural force.

But then, in the middle of the 20th century, a new medium arrived on the scene that was destined to reshape our culture at least as comprehensively as typography had done. Television broke the inky grip of print on our culture. (Earlier, radio had created its own revolution, particularly in the way people spent their spare time. But radio originally consisted of a combination of print-read-out, music, and plays and serials written for a non-visual medium, and none of this broke the grip of print on our culture in the way that television subsequently did.)

At first glance, the era of electronic, audio-visual communication may look like a throw back to tribal, village, pre-literate culture. Television tends to put the person and the message back together again; it blurs subjectivity and objectivity; it puts communication back into the public domain.

In *Other People's Words*, Hilary McPhee (2001) refers to the impact of this culture shift on the role and nature of language:

As the boundaries shift between media and as electronic transmission brings the promise of other ways of delivering words around the world, only certain kinds of words will 'travel'. The words themselves are already changing.

My children read books, but other things are at least as important to them. They watch many more films than I ever did. They play music and paint pictures and design animations and Web sites and print fabrics and rock climb by their fingertips and work sometimes on Aboriginal cattle stations or meditate in mountain retreats where books are considered distractions from the soul's search for the real thing. And my youngest son has a highly-intelligent best friend who proudly claims he has read only three books in his life: *Rumblefish*, *The Twits* and he can't remember the name of the third.

Television has brought us the illusion of shared experience — a pale shadow of the actual shared experience of village life. This is one reason why we have come to think of television, and now the Internet, as a way of connecting us to some vast 'global village'. When Sven Birkerts (1994) refers to 'the ersatz security of a vast lateral connectedness', the important word is 'ersatz'. The global village — at least as popularly conceived in mass media terms — is a huge hoax. Electronic linkages do not make a village. Village life is about relationships, interactions and interpersonal communication; it is about incidental contact and about the sharing of all the subtleties and nuances of each other's lives; it is also about the acceptance of mutual obligations.

Television and the Internet may look as if they create that kind of culture on the screen and in cyberspace, but I believe we sell the whole idea of human communication short when we allow ourselves to believe that there is a global village, or that the mere transfer of electronic data is the same thing as interpersonal communication.

When we send a message to someone, that person doesn't perceive what we say as a simple message. Rather, they will perform a highly creative act of synthesising all the messages contained in what we say, how we say it, where and when it is said, by whom it is said, and so on. Audiences never perceive a single message: they put together a complicated bundle of messages that they extract from the totality of the encounter. Every aspect of the encounter is, in effect, a message. (This is why the message in what you are trying to say can so easily be swamped by all the other messages in how/when/where you say it.)

The most effective medium of communication will always be the channel of person-to-person, face-to-face, one-to-one contact, if for no other reason than that it offers the other person the most richly nuanced, most varied range of messages from which to synthesise their impression of what we are trying to convey. A slightly less effective medium would be person-to-person, face-to-face contact in a small group. A larger lecture situation (like the one in which this chapter was originally delivered) is not bad, but effectiveness is greatly diminished by the distance between speaker and audience, by the constraints of formality and by the greater number of people present, making close interpersonal contact virtually impossible. (In settings such as churches we try to enhance the effectiveness of communication between members of a larger group by immersing them in a rich blend of experiential messages — symbolic behaviour such as kneeling, and the use of visual imagery, music, and ritualised personal participation. We also tend to elevate the speaker — in a pulpit — which reinforces the impression of authority.)

I find myself wanting to say that personal contact is the only *true* medium of communication: all the other so-called communication media are really channels for the mere transfer of data. Yes, personal contact is only about data transfer too, in one sense, but the data involved are so complex and subtle that we would be hard pressed to extricate them from the dynamic interplay involved in a face-to-face encounter. Some electronic media approximate some aspects of person-to-person communication, but they all operate on such a dramatically reduced 'bandwidth' that most of the potentiality of the encounter is stripped away before we even begin. Indirect and impersonal media can nevertheless be highly effective and efficient means of sending and receiving messages, especially when they are used as an adjunct to communication through the channel of an established personal relationship: the handwritten letter and the telephone conversation, for example, are rough approximations of personal contact when they are used between people who already know each other well; the same might be said for email or an exchange of text messages via a mobile phone (though words on a screen obviously contain less subtlety and richness than handwriting or the complex pitch, tone and pace of a familiar voice on the telephone).

But at least such things as email and text messaging occur on a one-to-one basis, and can be used to supplement personal contact. Communication via the *mass media* involves a process quite unlike the process of personal communication between two people in the same place at the same time. One of the key differences, of course, is our inability to give feedback and to create, together, a mutually-satisfying synthesis out of all the material each of us brings to that encounter. (Even the phone allows us to be more defensive, more disguised, more distanced, more 'phoney' than we can be when we are face-to-face.)

Print is the most distancing of all the indirect media of communication. People who, like me, have grown up in a culture dominated by print are inclined to treat literacy as the sovereign communication skill and to regard the printed word as having a status superior to all other forms of mass communication. We have been so enamoured of print, and of our own literacy skills, that we have lost sight of the paradox in a medium that appears to bring us together but actually keeps us apart. (This is the paradox of *all* the mechanical and electronic media: they keep us at a safe distance from each other.)

So the 'vast lateral connectedness' to which Birkerts refers is a very different thing from the experience of living in community with other people. Younger people who have grown up in a culture dominated by television and the Internet may be inclined to confuse this sense of electronic connectedness with human communication, though the smartest of them know, almost intuitively, that these media are mere adjuncts to the real thing, and the 'real thing' is personal relationships.

'HUMAN ABSENCE': A NEW WAY OF THINKING ABOUT COMMUNICATION?

Writing about new forms of identity construction, Karen Cerulo (1997) says this:

> New communication technologies have freed interaction from the requirements of physical co-presence. These technologies have expanded the array of generalised others contributing to the construction of the self. [The phenomenon of] the generalised other in a milieu devoid of place has created [the possibility of] the establishment of 'communities of the mind' and the negotiation of co-present and cyberspace identities.

Cerulo is referring to the emerging possibility of a sense of connection with others that no longer depends upon the central requirement of traditional, interpersonal communication: *being there*.

Such a prospect seems daunting to older people whose ideas about communication, relationships and identity have been rooted in the culture of face-to-face contact. Such people are inclined to regard the erosion of 'human presence' as a blow, quite literally, to civilisation as we know it.

Clearly, the culture shift wrought by the electronic media revolution does pose a genuine threat to our cherished sense of what constitutes communication. Print did that too, by separating the person from the message, but at least print didn't shift our sense of time and space in *personal* encounters, the way the mobile phone and the Internet are now doing.

Sven Birkerts (1994) describes this process of separation and dislocation with a sense of foreboding:

At earlier stages of history, before the advent of the sense-extending technologies, human interactions were necessarily carried out face-to-face, presence-to-presence. Before the telephone and the megaphone, the farthest a voice could carry was the distance of a shout. We could say, then, that all human communication is founded in presence. There was originally no severance between the person and the communication.

The telephone obviously altered that. It eliminated the need for a spatial proximity while keeping the time link intact ... bring in the answering machine, the voicemail, and the time link is cut.

But the telephone and answering machine are only a small part of the picture. A comparable set of transformations has taken place on every front. Handwritten letters gave way to typed letters, which became word-processed letters, a great many of them structured in advance by the software. And now email chatter is making inroads on the tradition of paper, envelope and stamp. Photographs, home movies, camcorder records of the stages on life's way. Every set of technological advances, every extension of the senses, involves some distortion of the time-space axis — the here and now — that used to be the given.

And who can say what the effect of all these changes and enhancements will be?

Birkerts, like many other concerned observers of the electronic revolution, sees information technology as supplying the components out of which a new kind of 'human hive' is being built: he describes it as a kind of 'amniotic environment of impulses' — a condition of connectedness that seems, in some disturbing way, to be spurious. Birkerts fantasises about how strange and exhilarating it will feel, in the future, for a person to stand momentarily free of all this connectedness — just as it now feels for a city dweller to be out in the country at night and see a sky full of stars.

This sense of connectedness, which is so different from the traditional view of interpersonal communication, has led some people to perceive a new kind of intimacy, unique to cyberspace ('falling in love on the Net' is a phrase that seems to make sense to some people), while others have entered that space with a new sense of *dis*connectedness and irresponsibility. 'Flaming' has become the ultimate in depersonalised communication: the hurling of abuse across the planet to people you have never met, and never will meet, but who have become the targets of your unfocussed hostility.

The question arising from all such considerations is this: is 'human presence' necessary to the experience of human communication? When you ring someone who knows you, they recognise your voice and the phone call is clearly coming from you. If they know where you live or work, they can probably visualise the place you are calling from as well. But the mobile phone has begun to change that: yes, it's recognisably you on the line, but the first question the called party wants to ask is, 'Where are you?'. Then the answering machine breaks the time link: you hear my voice, and you might even 'answer', but you know it's not me in the true sense — not here and now, anyway.

Email adds another dimension to the puzzle: when you get an email that appears to be from me, it's probably from me, but can you be sure? There are some disturbing stories now being told about fake emails purportedly sent by one person, but actually sent by another. On the Net, forgery, deception and disguise have become easier than ever.

Once you move into a chat room, you could be anyone: one of the liberating aspects of Internet exchanges, for some people, is that they offer the opportunity to take on some character, some persona, some identity you've dreamed of becoming: now you can try it out, see how you go, and see what kind of response you get ... yet still remain safely hidden from your audience. This creates the possibility of a new sense of identity — a new *meaning* of 'identity' — which, for some Internet users, actually enhances and enriches their experience of 'being me': their fantasy life has entered into their contacts with others.

In Saul Bellow's 2000 novel, *Ravelstein*, the character Rakhmiel reflects on such matters:

> The challenge of modern freedom, or the combination of isolation and freedom which confronts you, is to make yourself up. The danger is that you may emerge from the process as a not-entirely-human creature.

That idea of 'the combination of isolation and freedom' is integral with the new sense of universal connectedness that the electronic media culture creates. The gradual but steady erosion of human presence creates a new cultural framework in which representations of reality will pass for reality; data exchange will pass for communication.

Marshall McLuhan (1964), the 60s high priest of popular culture, once told the story of a mother wheeling her baby in a pram through New York's Central Park. A passing pedestrian stopped to look in the pram and remarked, 'What a beautiful baby!'. The mother replied: 'Oh, that's nothing. You should see his picture'.

The danger inherent in our acceptance of 'human absence' is that we might begin to think of *mediated* information as being the same as — or even superior to — our three-dimensional, here-and-now reality.

Bill Gates of Microsoft once remarked that an exchange of emails is a great preparation for a meeting, and a great way of recording the outcome of a meeting, but it's no substitute for a meeting.

When information cascades over us in the quantity that it now does, we may begin to think that the quantity is, of itself, a virtue. Information is both cheap and plentiful in our information supermarket and many of us have developed an insatiable appetite for it. The electronic media revolution has already turned into the information revolution: we are adrift on a sea of information. We are creating the kind of culture where information is treated as if it's an intrinsically valuable commodity whether it is relevant to our lives or not, and whether we have the time to interpret it or not.

Once we fall for the idea that information is good for us, it's a short step to the conclusion that more of it must be even better. (Is this the latest form of materialism?) Information has become the new status symbol.

But information is not the pathway to happiness, enlightenment or wisdom, any more than material prosperity is the pathway to any of these things. We can easily be distracted from the quest for our life's meaning by our dalliances with data. I raised in the second chapter the possibility that we can become so preoccupied with the sending and receiving of messages — and so seduced by the machines that carry those messages — that we might allow ourselves to become personally isolated from each other even while embracing the illusion that we are wired, connected and 'in touch'.

The Pulitzer Prize-winning novelist, Richard Ford (1999), describes a personal dilemma arising from all this:

> People who know a lot about technology would like to console us with their faith that it's neutral, that tools won't change human nature. But how do they know? And what if they're wrong? Or right? What is human nature anyway, and why do we think it's so well settled in us that we can't louse it up by taking it for granted?
>
> Put simply, the pace of life feels morally dangerous to me. And what I wish for is not to stop or even to slow it, but to be able to experience my lived days as valuable days. We all just want to keep our heads above the waves, find some place to stand. If anything, that's our human nature.
>
> I don't have email. I'm not on the Internet. I don't have a cellphone or call waiting or even a beeper. And I'm not proud of it, since my fear, I guess, is that if someone can't find me using any or all of these means, they will conclude that, for technical reasons, I don't exist any more.

More and more of us are wondering whether, if we don't have email, we don't exist. And some of us who *have* email are beginning to wonder whether it is a kind of electronic quicksand that will suck us down and eventually smother us.

The CEO of a large Australian corporation, having been away from his office for a few days, returned to find 400 email messages awaiting him. He deleted the lot and cancelled his email address on the spot. That might seem like an extreme reaction but, more and more, people are beginning to realise that in a media-saturated society, drowning in information is a real possibility. Though we retain some control over our contact with the traditional mass media, the new micro-media (especially the mobile phone and email) are looming as potential masters, rather than servants: who hasn't complained about being accessible, 24 hours a day? (The theory, of course, was that *other people* would be accessible to *us*, 24 hours a day.)

It has become fashionable to speculate about the possibility of the Internet being harnessed in the service of democracy, uniting us all in a mega-forum in cyberspace, where we can participate, with equal access, in the democratic process and express our views on any issue. Could cyberspace offer a new kind of 'village green', perhaps, where we could all assemble at an electronic version of the public meeting, and where our leaders could take their cue from us at every turn?

Personally, I doubt it. The currency of the Internet is speed; its great strength is its immediacy. Speed and immediacy do not sound like the crucial ingredients in the kind of reflection, contemplation and careful discourse that ought to precede the reaching of decisions that affect us. Democracy, at its best, is about the *quality* of consultation, debate and decision-making: it is not about the quantity of data available, nor the speed with which decisions are made. The necessary inwardness — reflection — that produces good decision-making and brilliant leadership is unlikely to be enhanced by a web-based version of the public meeting; indeed, it is likely to be eroded. Isn't one of the current problems in our society that political leadership has become *too* constrained by public opinion research? Aren't our leaders already in danger of becoming the puppets of public opinion? Isn't leadership about creativity, boldness, vision and moral courage, rather than the analysis of opinion poll data? (Don't get me wrong: I'm all in favour of politicians using public opinion research, but only as an *input* to the decision-making process, not as a substitute for it.)

Some people have also seized upon the information technology revolution as a potential breakthrough in education (just as others have seen it as 'the answer' to religious evangelism). But listen to Neil Postman, in *Amusing Ourselves to Death* (1985):

> We delude ourselves if we believe that most everything a teacher normally does can be replicated with greater efficiency by a micro-computer. Perhaps some things can, but there is always the question, 'What is lost in the translation?' The answer may even be: everything that is significant about education.

'Human absence' may hardly matter when a transfer of data is purely administrative, technical or trivial. But where people are seeking an encounter with each other, or where a purpose as serious as education is being served, Postman views 'human absence' as an insurmountable problem. If the personal relationship is integral with the process involved, then how will the process proceed in the absence of the relationship, and how will the relationship exist in the absence of the people?

THE TRIBAL GENERATION: TURNING OUT BETTER THAN WE FEARED?

People like Birkerts, Postman and Ford represent a particular generation (the generation to which I also belong): that generation's passion for print — especially for literature — may blind it to a rather surprising reality about the rising generation of young people who have been immersed, from birth, in the warm bath of electronically-transmitted information.

Those of us who have witnessed the communication revolution — from the outside, as it were — are inclined to ask ourselves how this or that medium might be changing our lives, and to explore whether we like the changes. Those are legitimate questions. But for young people born into the present culture, such questions scarcely arise. Having been *in* the new media culture all their lives, the things that seem revolutionary and even disturbing to their parents and grandparents are simply the air they breathe. They are dealing with the new world of the media in their own way, because it's the only world they've ever known.

Some young people have virtually abandoned themselves to the new forms of electronic media — especially the Internet — and, in the process, become disconnected and even alienated from their families and local communities. But my observation of the members of the rising generation is that they are not, generally speaking, being victimised by their exposure to the mass media nor, indeed, by their ready use of the Internet.

Yes, their vocabulary might be shrinking and they might be creating new styles of media-based language, but all that says is that language, in their hands, is living and changing, as it always does. From the perspective of the past, language always seems to be in the process of degenerating: from the perspective of the present, language is serving us as it always has. Perhaps we need to remind ourselves that the evolutionary miracle of the human species is not new media technology but language itself.

But the most striking phenomenon to observe among the vast majority of members of the rising generation is the extent to which they are determined to connect with each other *in a personal way*, almost as if they have intuitively known that it would be hazardous to rely on media information to sustain them.

This is a more tribal generation of young Australians than we have ever seen. They hunt in packs; they are herd animals; they have rediscovered 'the gang' in a way that is strongly reminiscent of ancient tribal cultures.

They make extensive use of mobile phones, email, Internet chat rooms and, of course, the mass media. But what is the most important thing in their lives? The answer, for most of them, comes without hesitation: *my friends*. They appear to have recognised that their most precious resource for coping with life in an inherently unstable, unpredictable and insecure world is the resource of *each other*.

Their parents remark that this is a generation that 'beeps and hums', yet a great deal of the beeping and humming arises from their use of electronic communication technology to allow them to keep in touch *with each other*.

Their parents worried that their children's use of the media — both mass and micro — would isolate them from each other, and that is a problem for a small minority. But the opposite trend seems to be emerging for the majority: young people spend time with each other all day at school and then, as soon as they get home, ring each other up to raise important questions like 'Where are you?' or 'Who are you with?'. They sit at their computers, late into the night, communicating with each other in chat rooms that, theoretically, could put them in touch with people all over the world but that are often used like local area networks (which have the great advantage of being more private than phone conversations when parents are lurking).

It's starting to look as if the effect of being saturated by media is not as dire as we might have feared: rather than turning young people inwards, and away from each other, their facility with ever-more sophisticated media technology seems to have turned them towards each other. The members of this generation hug each other at the drop of a hat. They talk about their friends as though they are a support group they will have for life. They seem more inclined to 'look out for each other' than their parents' generation. There is a level of commitment to 'my group' that offers a rather encouraging signpost to our future.

Not all members of this generation are inclined to connect with the herd: it goes without saying that, even in a tribal culture, not everyone feels 'tribal'. As in every generation, there are isolates, hermits, shy introverts and people who will go to almost any lengths to avoid social contact. For such people in the rising generation of young Australians, the Internet is an unimaginable boon. In earlier generations, they might have buried themselves in books or spent endless hours on a CB radio network or, before that, hunched over a crystal set. Today, they can connect and interact with greater facility; they can find 'playmates' on the Net; they can hide *and* connect, both at once.

As the children of the new media age emerge into adulthood, they do appear, in some ways, like the products of ancient tribal and village cultures. They have a sense of identity that is powerfully attached to the idea of belonging to their group. They derive emotional security from this sense of belonging — sometimes even more than from their sense of connection with a nuclear family. This is partly because the nuclear family is in some disarray under the influence of a high divorce rate, and partly because the plummeting birthrate means that many nuclear families are too small to offer the safety and security of the traditional 'herd'.

These are not like the rebellious, iconoclastic gangs of the 1950s and 1960s — the bodgies, widgies, mods and rockers. Nor are they the simple peer groups of the past, forcing conformity on their members. Today's herds are more like surrogate extended families, offering their members the emotional support of being accepted and taken seriously. There are plenty of aspects of modern society that create the demand for such support: could one of them be the need to both compensate for and protect themselves from the potentially isolating effect of excessive media consumption?

The connectedness of young people will affect everything from their voting patterns to their behaviour at work. Employers will need to realise they are dealing, from now on, with young employees who are disposed to rate their personal relationships more highly than their jobs, when it comes to creating and maintaining a desirable lifestyle. They will also have to adjust to young employees who naturally gravitate to teams and groups, and who are less comfortable with the idea of working alone.

Politicians are already facing the fact that the new social tribes are more important to young voters than the old political tribes.

Voting patterns will be less stable than they used to be as members of the rising generation of voters exercise their well known preference for keeping their options open (another response to the instability of life in an uncertain, swiftly-changing world), and as they seek information about candidates and policies from a wider range of sources than the traditional, official media of political campaigning. 'Underground' political activity is part of the emerging counter-culture of cyberspace: its effect can be seen in the willingness of many young voters to weigh up all kinds of information — reputable and disreputable, reliable and unreliable — in making their judgments about political, economic and social issues. As Neuman (1991) put it, 'an interconnected network of audio, video and electronic text communication ... will blur the distinction between interpersonal and mass communication and between public and private communications'.

Marketing organisations and advertising agencies are becoming attuned to the new kind of consumer being shaped by the information revolution — more media-savvy, more sceptical about brands, more interested in non-material values, unimpressed by traditional brand values that aren't backed up by contemporary performance, more prepared to seek deals and opportunities beyond the conventional channels of advertising and retailing, more inclined to trust their own and each others' judgments rather than conventional wisdom. The so-called 'de-massification' of mass markets appears to be generating an increasing tolerance of diversity as tribes learn to accommodate their members' eccentricities and individual differences with the kind of unconditional support traditionally associated with family groups. 'That's cool' has become the new declaration of acceptance of an almost infinite range of personal preferences.

The church, too, will need to take account of the culture shift now taking place. From the church's point of view, it looks as if the romance with modern mass-media technology was a distraction. Yes, church leaders sometimes need to use the mass media to express a point of view to society-at-large, or to enrich community debate about a particular issue. (This is especially true when the community is yearning for a sense of vision and for moral leadership.) Yes, a website can be useful as an electronic noticeboard, and yes, the church — like business and politics — needs to closely observe how the media are shaping and reshaping our society … but that is only because we can't respond to a society we don't understand.

A signpost to the church's best response to the media can be found in the attitudes of this emerging generation of young people: *personal* contact, *personal* relationships and a strong sense of community are how they — and all of us — will survive and prosper in an unstable world.

For much of the second half of the 20th century, the church was in danger of falling victim to a kind of media idolatry — worshipping the media as the new cultural paradigm, too easily blaming the media for the ills of society and, indeed, for the problems of the church itself — and failing to recognise that the *primary* Christian challenge remains, as ever, the challenge of responding, on the ground, to the needs of a wounded society.

When members of the rising generation say 'let's connect' they sometimes mean 'let's shift a bit of data to and fro', but mostly they mean 'let's *connect*'.

It's easy to accuse them of being mistaken about what constitutes connection, but they might well have found a way of incorporating electronic linkages into a new and larger sense of connection than older people might appreciate.

I didn't expect to be saying such things about this media-saturated generation, but it's what the research keeps telling me (Mackay, 1997), and it offers some hope for the future direction of our society. Many young people are using the media creatively and constructively and, if they sometimes want to 'veg out' in front of the television or on the Net, how different is that from the rest of us? (Why should veging out with television or on the Net be regarded as inferior to veging out with an undemanding book? The once common parental injunction to 'turn off the television and go and read a book' was only ever a symptom of attitudes formed by the print culture.) Perhaps young people know, intuitively, something the rest of us have been slow to acknowledge: that television is, in the end, primarily a harmless pastime, and that the possibility of new electronic linkages offered by the Internet, apart from being a source of undreamed-of quantities of data, offers us a way of augmenting and enhancing, but never replacing, our personal relationships.

POSTSCRIPT: THE 'BIG BROTHER' PHENOMENON*

There's been a significant change in the behaviour of the Australian TV audience over the past six or seven years. In the mid-1990s, the top 10 programs were dominated by news and current affairs, as Australians engaged with a heavy agenda of social, economic and political issues commanding their attention: globalisation, foreign investment, youth unemployment, Aboriginal reconciliation, the republic, tax reform, population policy, etcetera, etcetera.

By the turn of the century, Australians were becoming fatigued by this agenda: on the evidence of social research — but also on the evidence of TV ratings — it looked as if we were beginning to disengage from 'the big picture' and turn our attention to more personal, domestic, local matters. In the top 10 TV programs, news and current affairs were being replaced by so-called 'lifestyle' programs — home decorating, backyard renovating, cooking, holidays — as well as the traditional 'soaps' and standard escapist fare of comedy, romance and violence.

It was as though Australians had decided the items on the big agenda were beyond their control; they wanted to narrow the focus and turn it inwards, concentrating on an agenda within their control: 'We can't do anything about the impact of globalisation on the company we work for, but we can decide what video we'll rent tonight; we can't make Aboriginal reconciliation happen, but we can decide what school to send the children to next year; we can't work out what kind of republic we should be, but we can work out where to go for the school holidays'.

The change in our TV viewing behaviour signals a period of self-absorption, self-indulgence and self-centredness. It is reflected in our renewed enthusiasm for 'retail therapy', our declining sense of compassion towards the poor and disadvantaged — including refugees — and our growing propensity to retreat into prejudice (Mackay, 2001).

The behaviour of the media audience is a particularly useful source of insight into a society's attitudes, values and interests. One of the striking things about media consumers is their voracious appetite for programs and news stories that portray the extremes of human behaviour — murders, rapes, massacres, accidents, disasters — both natural and man-made. People love to be taken to the very edge of what it means to be human; to explore the limits of human experience and behaviour — especially bad behaviour (which helps to explain the enduring success of the world's sensationalist tabloid press).

So perhaps it isn't surprising to find that 'Big Brother' dominated the TV ratings for several weeks in 2001. Here was a program that responded to our need to narrow the focus and disengage from 'the big picture'; it also took us, in its own bizarre way, on an exploration of some of the extremes of human behaviour.

Even greater than the tosh served up in the program itself was the tosh written about it by media commentators — both its supporters and its detractors. To see such serious attention being given to the program, you'd think 'Big Brother' was a media event of great cultural significance, rather than a slickly-packaged and cleverly-branded international TV ratings machine.

Yes, there are some cultural lessons to be learned from the sudden spate of media voyeurism represented by programs such as 'Big

Brother' and 'Survivor', and by the proliferation of website cameras that track the private lives of individuals who have chosen to brush their teeth for all the world to see. But these are lessons about the nature of boredom in a media-saturated society — not about the moral decline of Western civilisation.

In examining the phenomenon of 'Big Brother', we must begin by killing off the notion that such programs can properly be called 'reality' television. Offhand, can you think of anything *less* real than putting a group of strangers together in a house, or on an island, pointing TV cameras at them, and asking them — or members of the viewing audience — to work out who should be thrown out first?

In fact, 'Big Brother' is an example of total *unreality*, but it does reflect our current interest in events taking place within a closely confined, domestic environment. Watching the interaction between human 'rats' in some kind of experimental laboratory is just the kind of thing that would appeal to an audience trying to insulate itself from important political and economic questions.

But does 'Big Brother' represent a new genre in TV programming? While 'Big Brother' owes a debt to the original 'Candid Camera' program, its participants' prolonged exposure to a contrived situation, plus the audience's involvement in picking a winner, puts it in a slightly different category.

So is it a sitcom? Not quite … too much 'sit' and not enough 'com', I'm afraid. (Comedy can't be left to chance: it needs skillful writers and directors who control every nuance of dialogue and gesture — creating situations is the easy part.)

A soap opera, perhaps? Close, but the plotline is too simple and the actors too lacking in artifice to sustain the year-in, year-out loyalty of audiences for a 'Neighbours' or a 'Blue Heelers'.

But that's an important clue: this is like a form of plotless drama — a program where you set up a situation, explain the rules, and then let the participants rip. Does that remind you of something? It should: another example of plotless drama is sport.

Like sport, 'Big Brother' reveals people's character, rather than building it. Like sport, its simplicity appeals to our most primitive competitive and survival instincts. Like sport, it evokes sudden and intense emotions that dissipate quickly: unlike art, it's over when it's over.

So that seems to be the answer. 'Big Brother' is really a new form of professional sport but with the added appeal for the audience of being able to participate in the umpiring. At many professional sporting contests, spectators already try to do that, but 'Big Brother' formalises it by inviting the audience into the decision-making process. No wonder Australians devoured it so voraciously.

---

*This is a transcript of an informal talk given to the students of New College at a dinner held on 4 September 2001, during the week of the New College Lectures. The talk was based on a column that appeared in the *Sydney Morning Herald*, the *Age* and the *West Australian* on 4 August 2001.

BIBLIOGRAPHY

ACNielsen (2001) *Australian TV Trends.* ACNielsen Media International, Sydney.
Baker, Richard (2001) 'Turn off the TV to pacify your little bully'. *Sydney Morning Herald,* 16 January, p 3.
Bandura, A & McDonald, FJ (1963) 'The influence of social reinforcement on the behaviour of models in shaping children's moral judgement'. *Journal of Personality and Social Psychology* 76: 274–81.
Bandura, A (1973) *Aggression: A Social Learning Analysis.* Prentice-Hall, Englewood Cliffs, United States.
Benn, Stanley I (1967) 'Freedom and persuasion'. *Australian Journal of Philosophy* 45(3).
Birkerts, Sven (1994) *The Gutenberg Elegies.* Faber and Faber, London.
Burrett, Tony (1990) 'Industry mobilises to repel the anti-alcohol attack'. *Ad News* 21 September.
Cerulo, Karen (1997) 'Identity construction: New issues, new identities'. *Annual Review of Sociology* 23: 385–409.
Cialdini, Robert B (1984) *Influence: The Psychology of Persuasion.* The Business Library, Melbourne.
Dale, David (2001) 'Now you see it, now you don't'. *Sydney Morning Herald* 28–29 April, p 27.
Dawkins, Richard (1989) *The Selfish Gene.* Oxford University Press, Oxford.

De Bono, Edward (1971) *New Think*. Avon, New York.
Felson, Richard B (1996) 'Media effects on violent behaviour'. *Annual Review of Sociology* 22: 103–28
Feshback, S & Singer, R (1971) *Television and Aggression*. Jossey Bass, San Francisco.
Ford, Richard (1999) 'A Waste of Time', *New York Times*, reprinted in the *Sydney Morning Herald: Spectrum*, 23 January.
Fox, JA & Pierce, G (1994) 'American killers are getting younger'. *USA Today* 24–26 January.
Freedman, JL (1984) 'Effects of television violence on aggressiveness'. *Psychological Bulletin* 96: 227–46.
Horin, Adele (2001) 'Care for kids with greater quality time'. *Sydney Morning Herald* 28–29 April.
Joy, LA, Kimball, MM & Zaback, ML (1986) 'Television and children's aggressive behaviour'. In TM Williams (ed.) *The Impact of Television: A Natural Experiment in Three Communities*. Academic Press, New York, pp 303–60.
Joyce, Timothy (1980) 'What do we know about how advertising works?' In S Broadbent (ed.) *Market Researchers Look at Advertising*. Sigmatext, London.
Klapper, Joseph T (1960) *The Effects of Mass Communication*. Free Press, New York.
Knickerbocker, HR (1939) 'Diagnosing the dictators'. *Hearst's International-Cosmopolitan*. Reprinted in William McGuire and RFC Hull (eds) (1980) *CG Jung Speaking: Interviews and Encounters*. Picador, London.
LaFree, Gary (1999) 'Declining crime rates in the 1990s: Predicting crime booms and busts'. *Annual Review of Sociology* 25: 145–68.
Leyens, JP, Camino, L, Parke, RD & Berkowitz, L (1975) 'Effects of movie violence on aggression in a field setting as a function of group dominance and cohesion'. *Journal of Personality and Social Psychology* 32: 346–60.
Mackay, Hugh (1983) 'Children and television'. *The Mackay Report*: June, Mackay Research, Bathurst.
—— (1986) 'Television'. *The Mackay Report*: November, Mackay Research, Bathurst.
—— (1991) 'Our evolving relationship with television'. *The Mackay Report*: October, Mackay Research, Sydney.
—— (1994) *Why Don't People Listen?* Pan, Sydney. (Republished in 1998 as *The Good Listener*. Macmillan, Sydney.)
—— (1997) *Generations*. Macmillan, Sydney.

—— (2001) 'Mind & mood'. *The Mackay Report*: July, Mackay Research, Sydney.
McLuhan, Marshall (1962) *The Gutenberg Galaxy*. Routledge & Kegan Paul, London.
—— (1964) *Understanding Media: The Extensions of Man*. Routledge & Kegan Paul, London.
McPhee, Hilary (2001) *Other People's Words*. Picador, Sydney.
Messner, SF (1986) 'Television violence and violent crime'. *Social Problems* 33: 218–35.
Messner, SF & Blau, JR (1987) 'Routine leisure activities and rates of crime: A macro-level analysis'. *Social Forces* 65: 1035–52.
Milgram, Stanley (1963) 'Behavioural study of obedience'. *Journal of Abnormal and Social Psychology* 67: 371–78.
Milgram, Stanley & Shotland, RL (1973) *Television and Antisocial Behaviour*. Academic Press, New York.
Miller, Jonathan (1971) *McLuhan*. Fontana, London.
Neill, Rosemary (2001) 'O'Shane furore underlines hypocrisy'. *The Australian*, 22 June, p 15.
Neuman, WR (1991) *The Future of the Mass Audience*. Cambridge University Press, New York.
Packard, Vance (1957) *The Hidden Persuaders*. Penguin, Harmondsworth, Middlesex, United Kingdom.
Parke, RD, Berkowitz, L, Leyens, JP, West & S, Sebastian, RJ (1977) 'Some effects of violent and non-violent movies on the behaviour of juvenile delinquents'. *Advances in Experimental Social Psychology* 10: 135–72.
Schramm, WJ, Lyle, J & Parker, E (1960) *Television in the Lives of Our Children*. Stanford University Press, California.
Schwartz, Tony (1973) *The Responsive Chord*. Anchor Press/Doubleday, New York.
Sweeney Research (2001) *Eye on Australia*. Information Australia, Melbourne.
Tedeschi, JT, Smith, RB III, Brown, RC Jr (1974) 'A reinterpretation of research on aggression'. *Psychological Bulletin* 89: 540–63.
Wiegman, O, Kuttschreuter & M, Baarda, B (1992) 'A longitudinal study of the effects of television viewing on aggressive and antisocial behaviours'. *British Journal of Social Psychology* 31: 147–64.
Wilson, James Q (1993) *The Moral Sense*. Free Press, New York.
Winn, Marie (1978) *The Plug-in Drug*. Viking/Bantam, New York.
Wood, W, Wong, FY & Chachere, JG (1991) 'Effects of media violence on viewers' aggression in unconstrained social interaction'. *Psychological Bulletin* 109: 371–83.